THE SERENDIPITY MACHINE

THE SERENDIPITY MACHINE

A voyage of discovery through the unexpected world of computers

DAVID GREEN

First published in 2004

Allen & Unwin
83 Alexander Street
Crows Nest NSW 2065
Australia
Phone: (61 2) 8425 0100
Fax: (61 2) 9906 2218
Email: info@allenandunwin.com
Web: www.allenandunwin.com

National Library of Australia
Cataloguing-in-publication entry:

Green, David G., 1949– .
The serendipity machine : a voyage of discovery through the unexpected world of computers.

Includes index.
ISBN 1 86508 655 X.

1. Computers - Technological innovations. 2. Information technology. 3. Technological complexity. I. Title.

004

Typeset in 11/14 pt Adobe Garamond by Midland Typesetters
Printed by CMO Image Printing Enterprise, Singapore

10 9 8 7 6 5 4 3 2 1

CONTENTS

PREFACE

The central theme of this book is the information revolution. In academic publications I have argued that informatics—the application of information technology to solve real-world problems—is emerging as the new paradigm for understanding and managing complexity in the real world. This paradigm is beginning to make itself felt in many practical ways, particularly in research, in business and in government.

In this book, I attempt to explain some practical aspects of this revolution. This will mean looking at controversial social issues, including data mining as a threat to privacy, the contribution of communications to globalisation, biotechnology and designer humans. The best-known sign of the information revolution has been the explosive growth of the Internet, which is discussed in several chapters.

One aim of this book is to help general readers to get beyond the hype surrounding some of the above areas and to understand the key ideas, issues and implications. The intent is not to provide all the technical details, but to explain the ideas and motives that underlie the technology. To do this, I have drawn extensively on my own experience and background, which includes research and development in many of the areas discussed.

Complexity is one of the hallmarks of the modern world. Most of us do not realise that we devote a lot of time and effort to reducing complexity in our lives. Society is full of conventions, institutions and other devices designed to help to simplify things for us. Many of the hottest social issues, such as those mentioned above, arise when these devices break down and complex, unexpected results emerge.

The idea of 'serendipity' provides a thread that runs throughout this book. The term denotes 'accidental' or 'fortunate' discoveries. My contention is that chance events are intimately bound up with the information revolution. One of my aims is to show that complexity makes serendipity and its effects inevitable. These effects are very widespread: serendipity not only underlies discovery, but also many natural and artificial processes, including both beneficial and harmful events. In nature, it plays a role in such diverse processes as species extinction and the workings of the brain. In human fields of activity, it influences accidents, luck, creativity, comedy and stock market crashes, to name but a few. The serendipity effect also plays a role in new computing functions such as data mining, evolutionary programming and multi-agent systems.

Sometimes ignorance really is bliss. A visit to a library can be a dismal experience. The reason is not the décor, nor the librarians—they are always friendly and helpful. No, it is that mountain of information. Inevitably, it turns out that someone, somewhere has already thought of, written down and published every bright new idea that comes to you. There is no quicker way to kill a half-baked theory than to find that someone has been there before you. I found myself visiting libraries less and less—even when I did, I rarely risked browsing the shelves. Ignorance made it possible to get on with the job, unhindered by fear of being pipped at the post. This strategy proved to be the best course of action as, more often than not, all those other authors had really taken a different tack after all.

One of the depressing side effects of the Internet is that you can no longer get away with self-delusion. While I was researching this book, I submitted several potential titles to Google, a prominent search engine. Much to my surprise, and dismay, it turned out that a whole string of authors had used the terms already. It mattered not that several references were to my own work. Other authors had used them years before me.

As usual with any book, the author is indebted to many people. Above all others, of course, there is my family. My wife Yvonne and daughters Audrey and Hilary have all shown endless patience and forbearance in tolerating computers at the dinner table, arguments over chapter titles and endless lectures out of the blue on obscure topics, not to mention a hundred and one other minor irritations.

Friends and colleagues have also contributed in many ways, large and small. I thank those colleagues and friends who provided stimulating debate, as well as a sounding board for ideas. People such as Roger

Bradbury, Terry Bossomaier, David Cornforth, Mike Kirley, Nick Klomp, David Newth and Glyn Rimmington. And then there are those, such as Peter Whigham and his wife Desley Anthony at the University of Otago, who provided a quiet retreat from the pressures of teaching and administration during the early stages of editing, as well as providing useful feedback on drafts.

I should also thank Hewlett Packard for inventing the Jornada hand-held computer, which allowed me to write many thousands of words in unlikely places (places such as Café Design and Café Mountain, where I spent many hours getting high on coffee and writing while my family spent my royalties in advance); Qantas stewards, who served me countless drinks of rum and cola while I tapped out chapters on trans-oceanic flights; and Japan National Railways for their Shinkansen bullet train, in whose green car I spent hours trying to rattle off words as fast as the countryside flashed by.

Finally, and above all, I want to thank all of you who take the time in a busy world to read this book. I hope that you will gain as much from reading it as I have in writing it.

chapter 1

THE INFORMATION EXPLOSION

To err is human, but to really foul things up requires a computer.
Anonymous, *Farmers' Almanac*, 1978

Computers everywhere

It's a risky business to make predictions about the future of technology. In the 1940s, Tom Watson, then chairman of IBM Corporation, was asked about the future of computers. 'There is a need,' he replied, 'for perhaps five computers in the world.' Few predictions have ever been less accurate! Within 50 years, computers numbered in the tens of millions. By the year 2000, almost every business, every government department and every home in western countries had a computer of some sort. Computers, and more generally information technology, have become the norm.

Besides growing more numerous, computers have also grown more powerful. In 1955, the IBM STRETCH computer was heralded as a great leap in processing technology. It could perform around 5000 floating point operations per second. Nearly 50 years later, the world's fastest computer realised a maximum processing speed of about 39 Teraflops, an increase by a factor of nearly 8 billion.[1] Computer memory and storage capacity likewise have increased by several orders of magnitude.

Today's computers are also more compact than their predecessors. In 1971, I had to process the results from an experiment in batches because the local mainframe computer could not store all 1200 data records in its memory at one time. That computer occupied a large, air-conditioned hall and was run by teams of technicians. Thirty years later, I can slip into my pocket a machine that can easily store spreadsheets holding ten times the number of records that stumped my mainframe in the 1970s.

We live in a world that is flooded with information. Incredible advances in information technology during the second half of the twentieth century enormously increased society's ability to *gather* data, to *store* data, to *process* data and to *transmit* data. On the one hand, it has enabled us to tackle complexity directly by keeping track of literally every detail. On the other hand, the challenge of dealing with mountains of information has created a new form of complexity of its own.

This book is about the information explosion and the role that computers play in it. In later chapters, we will look at recent developments in computing that are aimed at coping with our information-rich environment.

What is the serendipity effect?

Traditionally, the word 'serendipity' means a fortunate accident that leads to an unexpected discovery. The eighteenth-century English novelist Horace Walpole coined the term, although its origins are in an old Persian fairy story *The Three Princes of Serendip* (modern day Sri Lanka), in which the heroes travel the world making fantastic and unexpected discoveries.

The history of science and technology is full of examples of serendipity. Christopher Columbus set off for China and discovered the New World instead. Galileo turned his telescope on Jupiter to see it better and discovered that it had moons. Wilhelm Konrad Röntgen discovered X-rays by accident during the course of his experiments with cathode rays. Alexander Fleming discovered penicillin in a dish that had accidentally been left near an open window. Edward Lorenz accidentally discovered chaos because his computer rounded off numbers in his weather simulations. In all of these cases, and in many, many others as well, people have made important discoveries while they were trying to do something else.

Combining lots of different pieces of data leads inevitably to serendipity, to unexpected discoveries. Some of these discoveries are trivial. But even if only one in a thousand combinations leads to some unexpected discovery, the potential number of discoveries—unexpected, serendipitous discoveries—is still enormous.

To recognise the above process, I coined the term *serendipity effect*. Combining new data with old is one of the most common sources of the serendipity effect. It is almost a truism that new technology inevitably yields new discoveries. This is why astronomers always want to build better telescopes and physicists always want to build bigger particle accelerators. As we shall see in Chapter 6, the same desire helps to drive the current trend for building enormous data warehouses.

In a nutshell, the serendipity effect is a process by which the interaction of many different factors makes a particular outcome inevitable. In the above examples, the outcome was an unexpected discovery. However, in its broadest sense, serendipity is not confined to scientific discovery, far from it. If we think of life as a journey then every day we are exploring the future, so all unexpected events are a form of serendipity. This means that serendipity need not be confined to happy events—unwelcome discoveries, unwelcome events are results of exactly the same process. We will look at some of these closely related phenomena in more detail in Chapters 2 and 4.

The central theme of this book is that the computer is an engine for creating serendipity—a *serendipity machine*. In later chapters, we will look at the many surprising effects, the serendipities, that have arisen as a result of the information explosion. The computer is a machine that handles information; in doing so, it is changing the way we do things. It is changing the world.

Data harvesting

One result of the explosion in computing power is an enormous increase in our ability to gather, store and use information. This is especially true wherever computers are used to control or monitor situations. For one example, to check on environmental conditions scientists used to have physically to go out into the field and take the measurements in person. Nowadays, the combination of compact modern computers and communications simplify the process. A recorder can sit out in the field for months constantly transmitting readings back to base.

When we think of a 'computer', most people probably think of a PC sitting on a desktop. However, computers today come in a vast multitude of shapes and forms. As computers grew smaller, it became practical to place them in all manner of devices. Today, we see computers inside watches and clocks, inside the control panel of cars, and in a host of home appliances, such as radios, videos, washing machines, cookers and sewing machines. It is far easier to program a digital computer with the logic needed to control a device than it is to design mechanical devices that achieve the same effect. Digital control makes many new functions possible, such as programming your CD player to play a selection of tracks in the order that you prefer. Likewise, computers have found many specialised uses in business, such as cash registers, barcode scanners, and robots in manufacturing plants.

Specialised computers make it possible to gather data. Supermarkets can not only scan your purchases and add up their cost, but at the same time keep a record of every transaction and automatically adjust stock inventories. Most companies maintain databases for staff, clients and products. Electronic accounting packages help small businesses to keep track of cash flow and to calculate tax automatically.

Not so many years ago, most people used to own bank passbooks in which the teller would record the details of every deposit or withdrawal. Today, there is no need for passbooks. Computers record every transaction and the bank sends out a statement every month.

Computers have made possible a rapid growth in the popularity of credit cards. Once they were the status symbols of travelling business executives. Today, plastic is rapidly replacing cash as the standard method of payment. Transactions involving cards are transmitted electronically to the agency involved. As the use of cards has increased, so too has the flow of data about consumer purchasing patterns.

The automation of many processes has made it possible to collect huge amounts of raw data. For example, satellites transmit images back to earth several times daily. These high-resolution images may each be a gigabyte in size, so every year agencies need to store terabytes of raw data.

Many professions have formed international organisations to compile essential data. Although each individual dataset, produced by an individual may be relatively small, the joint activity of thousands of people adds up to a lot of data. Astronomers, for instance, have used this approach to compile international databases containing star catalogues, comets and photographic sky surveys.

Our ability to store huge amounts of data has increased along with our capacity to compile data. During the 1980s, the mainframe computer that I used allocated around 250 000 bytes of storage capacity to each user. Today, the hard disk of the personal computer that I use has about 100 000 times the storage capacity of that mainframe. Put another way, I have on my desk the equivalent of some 25 000 books, as much as a small library. Mass storage devices are even more impressive. Data storage on high-end machines is currently measured in terabytes, and will soon enter the realm of petabytes.[2]

In the 1950s, the jukebox was one of the most popular devices among teenagers. It was essentially a box containing dozens of vinyl records, usually the latest hits. When you put your money in the slot, and pressed your selection, a robot arm would retrieve the selected recording from its sleeve, place it on a turntable and the tune would start. Many mass storage devices are like a digital jukebox, but instead of vinyl records, they usually contain tapes or CDs. There may be thousands of individual tapes and several can be online at the same time.

Devices with such vast capacities of data are solutions looking for a big problem. One role they have filled is to act as repositories for the mountains of data now being generated. These *data warehouses* are now common in many fields of activity.

The information deluge

For most of human history, information was scarce and expensive; now it is abundant and cheap. From that reversal stem many changes. The flow of information solves many problems, but it creates many more.

An inevitable effect of increasing information flow is increasing complexity. Take the case of a typical large office. Once, if you wanted to draw a matter to someone's attention, then you wrote a memo. If the matter was important enough, you might copy the memo to one or two others as well. In the electronic office, however, it is as easy to send a memo to hundreds of people as it is to send it to just one. So most staff within an office are often informed about, and expected to contribute to, discussion that formerly would have involved only the office head. This may not be such a good thing when they may be dragged into issues about the minutiae of every section of their organisation and become less able to focus on their own job.

Amidst the torrents of information, it is easy to suffer from information overload. Electronic mail, for instance, is one of the great benefits of the Internet. In hours or even minutes, people can now complete exchanges that would have been slow or even impossible only a few years ago. For example, when a colleague in Japan sent me a draft of a new article that he had written, I was able to get comments back to him in a matter of a few hours. Using snail mail, this exercise would have taken weeks.

The sheer volume of daily messages has risen rapidly. On a bad day, there can be several hundred to deal with. Simply sorting this much mail can take up to an hour, even without reading it! For most people, advertising and other junk mail now comprises the bulk of their in-tray. My mail system provides filters, which automatically sort some of the mail, but there are always large numbers of messages that have to be read and dealt with.

One of the advantages of the Internet is that people with similar interests can join mailing lists to exchange news and ideas. When mailing lists first started appearing I quickly joined up with several that covered my main areas of interest. For a couple of years I even managed a few mailing lists myself. Now it is exciting to be a member of a list and to join in discussions before a worldwide audience, but an active list can generate dozens of messages a day. Multiply that by several groups and you're looking at hundreds of messages. In desperation, I eventually withdrew from almost all of the mailing lists to which I had subscribed.

In recent years, the problem has escalated dramatically. Many companies now use email to circulate notices and other internal communications. Worst of all, the Internet is a very appealing medium for advertisers, who are sending increasing amounts of junk mail. They can transmit advertising to a large, target group for next to no cost. Sure most people will simply delete the message but the practice will yield enough sales to make it worthwhile. From the consumer's point of view, junk mail is becoming harder to get rid of. Unfortunately, advertisers are becoming increasingly adept at disguising their junk mail as legitimate messages.

The problem is that email is unconstrained 'push technology'. That is, the source actively pushes information to the user. In contrast, accessing information via the World Wide Web is passive technology: the user pulls down only what he or she wants to see. One of the nasty things that can happen on the Internet is 'spamming', or deluging someone with email. The growing tide of unsolicited email has almost the same effect.

Serious games

The rapid spread of computers is due, at least in part, to their ability to help us to cope with complexity. They enable us to keep track of thousands, even millions, of items. They also help us to make sense of complex relationships, and to manage enormous systems successfully.

Chicago, Illinois USA, Thursday 5:15 p.m.

Sitting inside the cockpit of a 747 airliner, Joe Montana is making a final approach to busy O'Hare Airport. His eyes scan the instruments. All is in order. Approach radar indicates that the plane is right in the zone where it should be. Joe is feeling relaxed. He is confident. He has made this landing several times before. Suddenly an alarm starts ringing. The right wing engine has caught fire. For an instant, instinct takes over and Joe starts to shut the engine down. Then he remembers that he is landing, not at cruise altitude. What is the procedure? During this moment of indecision, a light starts flashing on his left. The hydraulics have failed! Instinctively, he guns the engines, trying to abort the landing. The right engine overheats and blows apart in a red ball of flame. The plane slews about in a slow arc. Joe tries to correct, but he can no longer control the flaps. He watches helplessly as the horizon tips at a crazy angle. The last thing Joe sees is the ground looming up through the cockpit window.

The lights shut down and a voice comes over the intercom. 'Bad luck Joe, want to try again?' Joe is a trainee pilot and is sitting in a cockpit simulator. Although he has learned the basics of controlling a large airliner, he has a long way to go before the airline will let him anywhere near a plane. Even as a copilot, he will have to have been exposed to, and successfully resolved, dozens, if not hundreds, of simulated emergencies, both great and small.

Commercial airliners are expensive pieces of equipment. No airline wants to risk one at the hands of someone who is not thoroughly trained and prepared. For this reason, airline pilots spend many hours training in flight simulators. They familiarise themselves with every detail of their plane's controls and its performance characteristics, until it is all second nature to them. They also learn how to cope with a host of emergency situations.

More generally, by placing us in virtual worlds, computers allow us to explore, to test and to experiment with complex situations before we have to deal with them in real life. This is especially useful when we need to manage a system that we cannot test in real life. For instance, to plan for global warming, we cannot perform experiments, nor can we afford to wait and see what happens, for that would be too late. There are many similar cases. To manage biodiversity, we cannot cut down all the trees and see whether anything survives. We cannot crash planes into buildings to see whether they will remain standing. (The twin towers of New York's World Trade Center did not collapse because of a failure of design. The architects had allowed for small planes accidentally hitting the structures. However, no-one had ever conceived that terrorists might deliberately crash airliners with full fuel tanks into the buildings.)

In each case, we need to perform 'virtual experiments' with virtual worlds now, to test possible strategies and gauge their likely outcomes.

Flight simulation is by no means the only use of computers in the airline industry. Modern air travel is a complex business. Every day, tens of thousands of planes fly on regular schedules from one place to another. Keeping track of those planes is a complex problem. Airports need to cope with large numbers of arrivals and departures, so it is vital they know how much traffic to expect. Air traffic controllers have to route the planes safely and efficiently. Airlines have to keep track of passengers. When you visit a travel agent in London and book a flight from Miami to Rio De Janeiro, your flight details need to be transferred to the airline and to the relevant ticketing authorities in the United States. All of this activity would be impossible without modern computers and communication networks.

Another important contribution of computers has been to allow people to learn how to manage complex systems in practice. One feature of complex systems is that they often behave in unpredictable ways. However, by simulating them on a computer, people can play 'what if' games. By trial and error with different scenarios, they can discover the best ways of managing a system under the worst case.

The knowledge Web

In 1990, the Internet was an arcane technology used only by academics and the military. Ten years later, in 2000, the term 'dot com' had become the mantra of a new era in entertainment and commerce. Information is of little use unless it is delivered to the people who need it. The need to deliver data from supplier to user led to explosive growth of the Internet. During the 1990s, the World Wide Web was the chief driving force behind this growth and the ubiquity of computers means that many people now have the means of handling and displaying information.

The Internet has been growing in size at an exponential rate ever since the first multimedia browser was released at the end of 1992. According to figures from the *Internet Software Consortium*, by July 2000 the Internet had grown to include 93 047 785 host sites.[3] Most of these machines would have supported World Wide Web servers. However, the actual number of users of the Internet and Web would have been much higher, perhaps by as much as an order of magnitude. The growth of the Web has helped it to expand from just over 1 million hosts in 1993: a factor of 100 increase in just seven years.

The World Wide Web has been growing likewise. A study published in the journal *Nature* estimated the size of the Web in February 1999 to be 800 million pages, with a total size of about 15 terabytes of text.[4] About 83 per cent of pages contained commercial content; scientific content accounted for only 6 per cent of pages. Within a year the number of pages had grown to a staggering 1 570 000 000 pages, including over 29 terabytes of text and nearly 6 terabytes of images. In other words, the size had nearly doubled within a single year. By early 2001, the search engine Google claimed to index over 1.6 billion Web pages.[5]

chapter 2

THE SERENDIPITY EFFECT

Something will turn up.
Mr Micawber in Charles Dickens' *David Copperfield*

They've all seen Star Trek

There is nothing like the threat of imminent doom to focus the mind. The year was 1990 and it was my first day on the job as leader of a small team that was building a database of the Australian environment.[1] I had been there just long enough to sign on and find my office when we were plunged into a crisis. We were going to be closed down!

Why? Because to outsiders it seemed that we were not doing our job. The problem was that everyone had seen the TV series *Star Trek*. From the lowliest clerk all the way up to the minister, everyone knew that if you wanted the answer to a question then all you had to do was to say: *Computer, tell me all about . . .*

Seconds later the computer would spit out a complete and detailed answer. That is the way it works in sci-fi films, so that is the way it works in real life too. Right? Not being computer experts, most people didn't realise, that a lot of hard work goes into designing and building even a simple database. Everyone expected that in a matter of weeks we would provide them with a system with which they could ask any question at all, about any issue, anywhere, and get back immediate, detailed answers.

Almost no-one had any idea of the enormous difficulties involved in developing a system that would be capable of answering even simple queries, let alone the futuristic system that they expected.

Creating an environmental information system for an entire continent is no easy task. The sheer size of the problem is daunting. Australia is a vast continent—nearly 8 million square kilometres. It also has a rich and diverse range of native plants and animals. For example, there are over 20 000 plant species.[2] There are also enormous technical issues to overcome, not to mention the delicate human problem of persuading hundreds of agencies and authorities to share their jealously guarded data.

This was the crisis I faced on the first morning of my first day on the job. It was clear that my priority had to be to dispel the misconceptions that surrounded the system. By carefully explaining the enormity of the task, and showing that we had a plan to accomplish it, we were able to forestall our execution. Nevertheless, expectations remained high and at times we felt that we had to achieve the impossible in no time at all, and with next to no resources.

How do you create a comprehensive information system? The answer is you do not. Instead, you set up a system that tells people what they most need to know. To do this we focused on a number of crucial questions, such as 'Where is animal species *X* found?'; 'What forests occur on public lands?'. We also developed priorities, such as providing background information about any location, including maps of vegetation cover, climate, land use, location of rare and endangered species and large-scale satellite images. For priority regions, especially the heavily populated southeast of the continent, we set out to capture more detailed information, such as fine-scale satellite images. Now that we had a strategy in place, we set about collecting the necessary data and installing software to interpret it.

Enter serendipity

What we designed, and what I have described above was an information system that focused on a limited range of issues. We had certain urgent questions that we needed to answer, so we collected the necessary data and resources to answer those questions and those questions alone. In the course of all this data collecting, a curious phenomenon emerged. Each new set of data not only answered the questions we wanted to ask, but also questions that we had never thought about or planned for. For instance, data on species distribution, which was needed to describe different environments, could when combined with satellite images help explain seasonal colour changes. In other words, we were experiencing serendipity—accidental discoveries. As we added more and more data, the number of questions we could ask, and the number of discoveries we could make, increased so rapidly that it soon went off the scale.

To understand this, suppose that any two sets of data can be combined to answer one question. Datasets can be used alone or in combination, so with 2 datasets we could ask 3 questions. With 3 datasets, there are 6 combinations. From there, the number of combinations increases rapidly: with 10 datasets there are 55 combinations and with 100 sets there are 5050 combinations.[3] And in a data warehouse with (say) 1000 sets, there would be over 500 000 combinations.

In reality, the number of questions that can be asked with a given combination of data is far greater than the above figures might suggest. For example, the data for plant locations included records for hundreds of separate species. Any question you could ask about one species, could be asked about hundreds of others.

Whenever you combine lots of information, you get unexpected results, and this serendipity effect is one of the hallmarks of the information explosion. As we shall see below, it makes itself felt in many different contexts. In later chapters, we shall see that it has led to many new ways of using computers to solve problems but it can also deliver some nasty surprises.

Coincidence

One of the funniest questions asked of travellers runs like this:

> 'Hi, where do you come from?'
> 'Australia. I live in Sydney.'
> 'Oh really. My cousin's studying in Australia. He's in Perth. Do you know him?'

Now Perth lies almost 3000 kilometres west of Sydney, about as far as Moscow from London, or New York from Los Angeles. Nevertheless, people assume that because you are both in the same country you must have bumped into each other.

Fantastic coincidences of this order sometimes *do* occur. When I visited London for the first time, I thought I didn't know a soul in the entire country. Nevertheless, I bumped into a friend of mine while walking down the Strand. Any frequent traveller can recount coincidences of this kind, but the point to note is that such coincidences are not *specific*. Unlike a pre-arranged meeting, where you set a time and place to meet a specific person, you cannot predict exactly where, when or whom you will meet. But given enough chances, the odds are quite good that eventually you will meet someone, or come across someone who knows someone you know. Having lived in Canberra for many years, I have a lot of friends and acquaintances in that city. So any time I visit and am wandering around the streets, the chances are quite good that I will bump into someone I know. Likewise, if I make a new acquaintance, then as often as not it turns out that there is someone we know in common.

Serendipity is another word for coincidence. And as the above example shows, you can make serendipity occur simply by ramping up the number of chances of a random event occurring.

Meaningful coincidence

Closely related to the serendipity effect is the idea of 'meaningful coincidence'. A meaningful coincidence occurs when chance throws up a combination of events in which we can detect a relationship, for example:

- While you are thinking about a friend, he or she calls you on the telephone.
- You are reading a novel and come across the word 'balloon'; putting down your book you look out the window and see a child playing with a balloon; you then turn on the TV and at once see a news report about hot air balloons.

The danger with such cases is that we often falsely jump to the conclusion that the pattern we detect indicates a real relationship. If friends phone while you are thinking about them, it is not necessarily telepathy. Scientists refer to this problem as confusing 'correlation with causation'.

To make this difference clear, try the following simple experiment. Imagine that you have a coin and that you toss it five times. Each toss results in either a head (H) or a tail (T). The end outcome is a string of five results, each being either H or T. Now before reading any further, write down your imaginary set of result.

Finished? OK, now look at your results. Do you have a sequence of five heads (HHHHH) or five tails (TTTTT)? If not, do you have a sequence of four of a kind (HHHH or TTTT) in your results? If not, do you have a sequence of three of a kind?

I have carried out this experiment many times with many different audiences. Often I find that even in quite a large audience no-one produces a sequence of three of a kind in their imaginary experiment, let alone a sequence of four or five of a kind.

In the real world, a sequence of (say) three tails (TTT) is just as likely as two tails and a head (TTH). The problem is that we confuse randomness with lack of pattern. We assume that a random result is devoid of patterns. In our imagination we see a pair of tails and detect a pattern emerging. We falsely assume that only a coin toss that disrupts this pattern is random.

The result of the above confusion is that we tend to dream up results that are extremely biased. In all, there are 32 possible outcomes of the coin tossing experiment: HHHHH, HHHHT, HHHTT . . . TTTTT. In a truly random experiment, any one of these results is just as likely as any other result, whether or not we perceive any meaning in it. Two of the possible outcomes are sequences of five in a row: HHHHH and TTTTT. So the chances of five in a row are 2/32, that is 1/16. Likewise there are 6 outcomes of at least four in a row: HHHHH, HHHHT, THHHH, TTTTH, HTTTT, and TTTTT (with a probability factor of 3/16) and 16 outcomes of at least 3 in a row (a probability of 1/2). So when, as sometimes happens, no-one in an audience of 50 people produces a sequence of three in a row the odds of that happening by chance would be only $1/(2^{50})$ or about 1 in 1 000 000 000 000 000!

It is easy to see that reading meaning into coincidence may have advantages. If a rustling in the grass precedes a tiger attacking, then the safest course is to be wary of rustling grass in future. The danger with meaningful coincidences is that people tend to mistake them for real relationships. So if you have an accident shortly after a distant relative comes to visit, don't assume that that person jinxed you.

Meaningful coincidences occur all the time in many different guises. The phenomenon has long been known. Carl Jung, one of the pioneers of psychoanalysis, used the term *synchronicity* for 'meaningful coincidence'. In his book *The Roots of Coincidence*, Arthur Koestler showed how coincidences confuse the whole question of whether extrasensory perception (ESP) really occurs. For instance if you are thinking of a friend when she phones you up, then you tend to assume that your friend must have been reading your mind. At least, that is the inference people tend to make.

Here is another popular game involving meaningful coincidences that illustrates the serendipity effect in action. Suppose that you are chatting to a stranger at a party and in the course of the conversation, you get around to talking about birthdays. What are the chances that you both share the same birthday? Ignoring leap years, there are 365 possible days on which your companion's birthday could fall. Only one of those days is your birthday, so the chance of you sharing a birthday is just 1 in 365. This is pretty small.

OK, but what happens in a room full of people? Suppose we set out to find whether any pair of guests at the party shares the same birthday. What are the chances of that happening? Most people think that you need a really big group, say 200 or so, but it turns out that the chances are very good even in relatively small groups.[4] If there are more than 50 guests at the party, then it is close to certain that there will be at least one birthday match. Table 2.1 shows how rapidly the chances of finding a matching pair increase.

Horoscopes and mind readers

Whether wittingly or not the authors of horoscopes take advantage of meaningful coincidences in casting people's fortunes. Mention enough

Table 2.1 Probability that a group of given size includes a pair with matching birthday

Number of People	0	20	30	40	50	60
Probability (%)	11.6	40.6	69.7	88.2	96.5	99.2

common events in a vague enough way and you are sure to strike a common chord somewhere. For example, look at the following 'prediction':

> A day for meetings and perhaps for new beginnings, but beware of overdoing it today of all days.

Take the word 'meeting'. This could apply to anything from a business meeting to bumping into someone on the street. Almost everyone 'meets' someone in some context every day. A 'new beginning' could also have all sorts of meanings, whether starting a new job or wearing a new outfit. Likewise 'today of all days' could refer to (say) a birthday, a day of the week, or a special event.

Fortune-tellers have two things working in their favour. First, people look hard to find associations between a horoscope and their own situation. Second, they jump on correct predictions and ignore false ones.

We see other examples of meaningful coincidences at work in many kinds of *post hoc* interpretation. Perhaps the most famous fortune-teller in history was Nostradamus, who lived in France during the sixteenth century. His main legacy is a collection of prophecies expressed as quatrains, verses describing visions of future events. However, as with horoscopes, the prophecies are vague and, in most cases, their supposed meanings have become clear only after the event. As many sceptics have pointed out, they are so vague that they could be made to fit almost any scenario.

Post hoc interpretation of coincidences also occurs in current events. For instance, after the terrorist attack on New York on 11 September 2001, there was a flurry of reports about signs and portents of the event. The date (9/11 in US notation) is the same as the American emergency dialling number 911. Likewise, there were many reports and claims about hidden messages in the Bible.

Mind readers and mediums likewise exploit the serendipity effect. For example:

Mind reader:	'I see a man, a tall man.'
Subject:	'Oh yes. That would be my brother.'
Mind reader:	'He is speaking and at the same time holding something in his hand, perhaps a pen or microphone or a similar object. Is that music in the background?'
Subject:	'That must be his speech at my wedding.'
Mind reader:	'Yes, I do hear music. And your brother is trying to tell you something, something important.'
Subject:	'Oh my goodness, yes. He was telling us to drive carefully on our honeymoon.'

A skilled 'mind-reader' can go in this vein, coaxing more and more information out of the subject. At the end of the session the subjects go away convinced that the mind reader has read their thoughts. They fail to realise that they have supplied most of the details themselves. The performer exploits the fact that in their haste to seize on every valid association, subjects usually fail to notice that most of the performer's guesses are wrong.

Others exploit the serendipity effect in similar fashion. Faith healers, for instance, rely on the their subject's belief in order to create a placebo effect. A positive mental attitude can produce at least a temporary improvement, and sometimes even a permanent cure, in many conditions. People remember the successes and forget the failures, or explain them away by blaming the subject's lack of faith.

In all of these cases people *want* to believe. They judge the performer on their wins, not their losses. There is no criterion for failure, unlike science, where theories are tested not on their successes, but their failures.

Whether we believe in horoscopes or not, the sad fact is that we all tend to assess people and things in much the same blinkered fashion. If we like someone then we tend to remember every good thing they do, and ignore or excuse their shortcomings or any bad deeds they may do. In contrast, if we think ill of someone, then in our eyes they may be unable to do anything right.

Likewise, our attitudes to life colour our interpretation of experiences. How many of us 'always' find ourselves in the slow queue in the bank or supermarket, or stuck in the slow lane in heavy traffic? The truth is that just as often we land in the fast lane, but being in the fast lane is not annoying so we promptly forget about it.

The above biases are unwelcome by-products of the way we process incoming information. Our senses are constantly being bombarded with more data than our brains can possibly process. We cope by selectively ignoring most of what we see, hear and feel—perceiving only a simplified vision of the environment around us.

Making your own luck

'Don't wait for things to happen, make them happen!' A cranky schoolteacher taught me this lesson many years ago, when he found a group of students standing outside a locked classroom, waiting for someone to turn up with the key. It was one of the most valuable lessons I ever learned, and perhaps its most important application is in creating opportunities.

Some people are famous for their luck. Julius Caesar, for example, was renowned for the way in which events always seemed to go his way. However, being lucky is usually the result of a lot of hard work behind the scenes. If you are always making opportunities, then every so often you will be lucky. If you sit back, wait and hope for good luck to strike, then the chances are that it never will.

Take gambling, for instance. Most people are taken in by the chance of winning a fortune, no matter what the odds. If you buy one ticket in a multi-million dollar national draw, then your odds of winning the big prize are vanishingly small. Buy 100 tickets and they are still pretty small. On the other hand, there are always plenty of small competitions around where the odds of winning are quite good. Many companies use lucky draws to promote their products. You have to be in it to win it, as the old saying goes, and people who enter promotional competitions on a regular basis find that they often win quite substantial prizes.

The same principle—creating opportunities—also applies to many games. In the card game Patience, for instance, one common mistake is to pile up cards too quickly. A more successful strategy is to keep as many piles open as possible. This creates the most possible opportunities to move cards and thereby further openings. (We will see more examples of this kind in Chapter 3.)

Serendipity often plays a big hand in two of the most important events of anyone's life: finding a career, and finding a mate. For most young people in western countries, arranged marriages are now a thing of the past. The search to find a life partner is a preoccupation that lasts from puberty until marriage, and even beyond. As everyone knows, you will never meet your perfect match sitting at home alone. You have to be out and about meeting people. Now in the search to find Mr Right, or Miss Perfect, you are likely to encounter many boys who are not quite right, or girls who are not quite perfect. Even when you do find someone who makes your heart beat faster, you might lose out to rivals for your intended's affections. It follows, then, that the more opportunities you have for social contact, the better your chances of success. In former times, the social conventions ensured that occasions existed on which young people could see and be seen. Today, however, most young people cannot rely on either family or social events to parade potential mates before them. They have to take a more active role in the search. This may mean heading off to dances or parties every Saturday night, or spending many long lunches extending your circle of friends and contacts, or indulging in an endless lottery of meetings arranged by introduction agencies.

Complexity and the serendipity effect

The serendipity effect is a natural result of complexity. Its chief underlying source is what is known as *combinatorial complexity*, which occurs because there are always many more ways of combining and arranging a set of objects than there are objects themselves.

Take the DNA code, for instance. There just four bases: Adenine (A), Cytosine (C), Guanine (G) and Thymine (T). These constitute the genetic alphabet. But within each gene, these four symbols are grouped into *codons*, which are strings of three characters. (The name codon derives from the fact that they code for an amino acid.) Now, if you look at the string of three characters in a codon you will see that, because there are four bases, there are four possibilities for the first character in the string. Likewise, there are four possibilities for the second and for the third places, giving us $4 \times 4 \times 4$, or 64 possible strings of bases.[5] Strings of three bases are necessary because although there are only 16 (4×4) different pairs of bases, there are 20 amino acids in all. However, because triplets provide 64 combinations, there is much redundancy. In many cases, the third base in the sequence is irrelevant: the same amino acid results whatever base occurs in the third coding position. In addition, the sequences TAA, TAG and TGA are 'terminators' or stop codons. They tell the ribosomes, which read the code and make proteins, where the sequence ends.

The point of this example is that arrangements of just four characters lead to a much greater number of coding possibilities. Likewise, 26 symbols are enough to build all the thousands of words in the English language. For instance, there are 456 976 possible 4-letter strings of the 26 characters in the alphabet. If we restrict the list to strings that contain at least one vowel, then this number falls to 262 495. This is still an extremely rich range to choose from. For words of five characters, the number of possibilities rises to nearly eight million.

Combinations are also rich in possibilities. Suppose that we have a pool of 10 items. For simplicity let's use the letters A, B, C, D, E, F, G, H, I and J. Now if we put them in a hat and draw one out, then of course there are 10 possible results. However, if we pull out two, then there are 45 possible pairs of letters that we can draw from this pool:

AB, AC, AD, AE, AF, AG, AH, AI, AJ,
BC, BD, BE, BF, BG, BH, BI, BJ,
CD, CE, CF, CG, CH, CI, CJ,
DE, DF, DG, DH, DI, DJ,
EF, EG, EH, EI, EJ,
FG, FH, FI, FJ,
GH, GI, GJ,
HI, HJ,
IJ.

Notice here that we are ignoring the order in which they are drawn. As the size of the pool we use, and likewise the size of the group that we draw out, increases, then the number of possible combinations can be very large indeed. Table 2.2 shows the number of possible groups that we can select from a given pool of items.

Table 2.2 Number of possible groups of different size in a pool

	Group size			
Pool	*2*	*5*	*10*	*50*
10	45	252	1	–
100	4950	75 287 520	1.73×10^{13}	1.01×10^{29}
1000	499 500	8.25×10^{12}	2.63×10^{23}	$\sim 2 \times 10^{85}$

Notice how rapidly the number of possible selections grows. The later numbers grow so huge that we cannot write them down easily. For instance, the number written as 2×10^{85} in the table means 2 followed by 85 zeroes! These are truly astronomical numbers.

Deaths in the Prussian cavalry

Serendipity's chance nature is closely related to many other rare events. In 1898 a famous study of rare events was undertaken in Prussia by Professor von Bortkiewicz. He examined the tragic fate of 196 members of the Prussian cavalry who had been kicked to death by horses. Now being kicked to death by a horse is a rare event, even in the cavalry. Nevertheless, the Prussian military had kept careful records of every case in fourteen cavalry corps during the 20-year period from 1875 to 1894.

Von Bortkiewicz discovered that although there had been a high number of deaths in some years, the deaths were truly random. The accidents however, although highly unusual, and extremely rare, were none the less predictable. You could not predict whether any particular cavalry officer was going to die, nor which regiments would suffer losses of this kind. But you could predict how the total number of accidents would vary from year to year.

Von Bortkiewicz showed that a precise mathematical law governed the number of cavalry kicked to death by horses. This *Poisson distribution* was

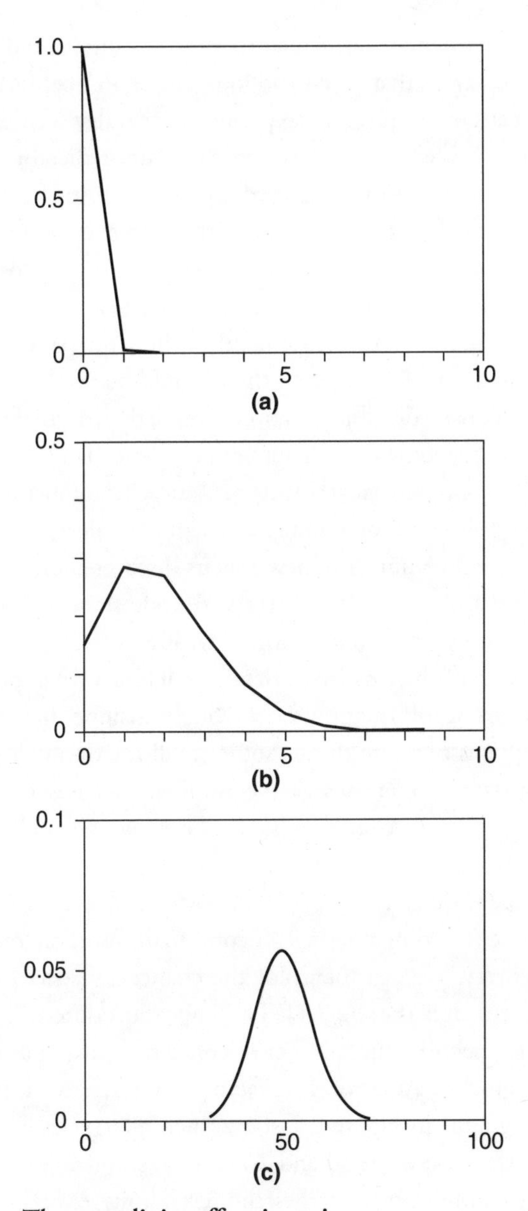

Figure 2.1 The serendipity effect in action
These three graphs show the probability of obtaining different numbers of successes for sets of sizes (a) 2, (b) 20 and (c) 100 respectively.

named after the mathematician Simeon-Denis Poisson, who discovered it in 1837. What it says is that when you look at the number of times some rare event happens over a given time period, say cavalry accidents for one year, then although each event is rare and unpredictable, the total numbers turn out to have an average frequency. Most of the time, the numbers will be very low, but occasionally they can shoot up. In one year, a single cavalry corps lost four riders. Concentrations like this often fool people into thinking that some deeper cause must be at work, but it is just the vagaries of chance. On the other hand, if the same pattern happened year after year in a single corps, then there would be reason for alarm.

Of course the Poisson distribution governs more than deaths at the hooves of cavalry horses. The same pattern applies to a great many everyday situations: the number of industrial accidents per year; the number of customers entering a shop each hour; the number of telephone calls taken by a receptionist per hour; or the number of new comets discovered each year.

The Poisson distribution also helps us to understand serendipity. We suppose that a rare event should happen with a certain, low frequency. Now as we have seen, chance discoveries often happen when people study new combinations of information. So we might assume that, on average, a discovery happens at a rate that is some small fraction of the cases we look at. But that rate increases as the number of cases increases. For instance, suppose that the chance of 'success' on any one 'trial' is 1 in 100. If you take a sample of 1000 cases, you could expect the overall number of successes to be about ten.

Let us suppose that only one in 100 combinations of factors produces a genuine discovery. We can then plot the chances of success for sets of factors of different sizes (Figure 2.1). The important difference is that, as we saw in the last section, the number of combinations of pairs of factors increases very quickly. A set of 100 factors gives rise to 4950 pairs of factors. As the graphs in Figure 2.1 show, such a large set will yield not one, but somewhere between 30 and 70 successes.

In summary, complexity underlies the serendipity effect. Unexpected discoveries within interacting databases represent just one kind of unpredictability. As we shall see later, serendipity translates into many contexts and into many phenomena.

chapter 3

DIVIDE AND RULE

Our life is frittered away by detail . . . Simplicity, simplicity, simplicity! . . . Simplify, simplify.
Henry David Thoreau[1]

Million dollar bugs

On the morning of 4 June 1996, the European Space Agency's rocket *Ariane 5* took off from the launch pad in Kourou, French Guinea. It was the rocket's maiden flight. At first, everything seemed normal but then, 36 seconds into the flight, and 3700 metres above the mangrove swamps, the launcher veered off its flight path, broke up and exploded.

What went wrong? According to the official accident report, the ultimate source of the failure was a bug in the software that performed the alignment of the rocket's strap-down inertial platform.[2] Unfortunately, the code failed to include a check to guard against this potential problem. Although an earlier rocket, the *Ariane 4*, had originally used the same computer program, the fault was never detected. On those previous flights, no part of the rocket's trajectory ever produced a condition that would cause an overflow. Unfortunately, the early part of *Ariane 5*'s trajectory differed from that of *Ariane 4*. This difference led to considerably higher values in horizontal velocity, which in turn led to the fatal overflow.

In short, a multi-million-dollar rocket was destroyed because of a tiny software fault. Software bugs have thwarted several other important space missions. In 1963, for instance, a simple typo—a missing bar symbol in a formula—led to NASA's *Mariner 1* probe, intended for the planet Venus, being destroyed by the range safety officer shortly after launch.

Problems of this kind are all too common. A natural reaction is to blame the engineers. How could they be so careless as to let a trivial error ruin an entire mission? But were they careless? The software for space missions often runs to hundreds of thousands of lines of code and here we see the serendipity effect at work again. The number of errors in a piece of software increases exponentially as the length of the program increases. This means that if you double the length of a program it is liable to contain four times as many errors. A program three times as long is likely to contain eight times as many errors, and so on. The trouble is that relatively few of these errors are simple typos. Instead, they tend to be subtle errors of logic that, as in the *Ariane* program, only show up under certain conditions. For example, one of the easier errors to spot is the command 'Divide the number Y by the number X'. This command works just fine almost all of the time, except when the value of the number X is somehow set to zero.

Software 'bugs' are not confined to esoteric systems like space flight control programs. Anyone who has ever used a computer has experienced the frustration of a program that crashes mysteriously. And then there are the more bizarre cases, such as people receiving final demands for $0 because someone forgot to write a bit of commonsense into a financial program.

Not surprisingly, computer programmers are not happy to hear such stories. It is truly frustrating to spend months designing, writing and testing an important piece of software only to see it crash the first time it is used. So over the years, programmers have devised ways to reduce the problem.

The most important idea that computer scientists use in order to eliminate errors is *modularity*, that is, break a big job down into self-contained parts (modules) that can be used again and again. The motivation here is not only to avoid errors but also to achieve greater efficiency. After all, anyone who is writing a program with a million lines of code needs to use every trick in the book just to get the job done. Modularity takes many different forms, but it runs all through the development of computer science, from its earliest days right up to the present.

In the very early days of computing, programs were written in machine code.[3] At this level of detail, mistakes were very easy to make, but programmers quickly discovered that they tended to use the same sequences of commands over and over again. This led them to invent the idea of *encapsulation*. That is, they saved often-used sequences of command and gave them names. Whenever they needed to use those commands, all they had to do was insert the name as a shorthand way of calling them up. This in turn led to the idea of macros and subroutines. A macro is a short sequence of commands that you store for future reference. By giving the macro a name, you can then call up those commands again at any time. This saves time. It also decreases the likelihood of making coding errors. The same idea applies to subroutines, which are self-contained sets of operations and are often used repeatedly within a program.

Now, if you can recall a set of commands by giving them a name, then why not write entire programs by calling up sets of commands by the names under which you store them? It did not take harassed computer programmers long to grasp this idea, and as soon as they did, high level computing languages were born.

That is not the end of the story, far from it. The quest for fast, reliable programming continues to this day. All high level computing languages include the ability to encapsulate commands in subroutines, functions and other structures.

Divide and rule

> *We can form a single united body, while the enemy must split up into fractions. Hence there will be a whole pitted against separate parts of a whole, which means that we shall be many to the enemy's few.*
>
> Sun Tzu[4]

Port Arthur, on Tasmania's southeast coast, was a notorious convict settlement during the early 1800s. Of the few prison buildings that remain intact, perhaps the most bizarre is the chapel. It looks like a normal church except for one detail: every seat is enclosed by a booth.

Although open at the front, the booths are so arranged that no prisoner could see any of his fellow inmates. The system had a cruel logic to it. It isolated each prisoner and helped to crush his spirit. It also prevented the inmates communicating with each other. If they could not communicate, then they could not plot and scheme to escape. Divide and rule.

From Julius Caesar to Machiavelli, political strategists through the ages have applied this potent political doctrine with great success. Set potential enemies at each other's throats and they will have neither the time nor the ability to oppose you. It was an especially effective tool of imperial powers. The British, for instance, used it to protect their colonies in Africa, India and North America.

More generally, the principle of divide and rule is the most common way of coping with potentially complex situations. Carve up a large, complex problem into small, simple problems and deal with them one by one. People use this approach routinely without even realising it. We see it applied everywhere, from business schedules to building a house.

How do you build a house? Newlyweds Kathryn and Mark want to build themselves a home. To save money they decide to be owner–builders. However this does not mean that they have any skills at building. All they do is divide up the work and coordinate things, farming each stage of the actual work out to professionals. They hire the legal firm of Speed and Loophole to handle the titles and contracts. They consult the architects Nifty Planners Inc. to draft up their home plans. Then they engage a string of contractors to handle different stages of the construction. Bill Woods and Co. provide the carpenters. The Clay brothers do the bricklaying and masonry. Joe Waters installs the plumbing and Eddie Powers installs the electrical cables and fittings. Just as a computer programmer carves up a large piece of software into compact modules, so our homebuilders divided one large project into several smaller, self-contained jobs.

The idea of modules is deeply embedded in our thinking. Take numbers, for example. The idea of building blocks in arithmetic probably began in prehistoric times with the custom of using fingers for counting. Having ticked off a set of objects with the fingers of one hand, you could refer to that set collectively as (say) a 'hand' or 'fist'. The person doing the counting could then enumerate larger sets by counting off the number of fists present. People still adopt this idea when they make a mark for each object and score a line through each set of five marks. The ultimate expression of this approach to counting is the decimal place system in which each digit represents entire sets of different size. The number 452, for instance, means 4 sets of 100, plus 5 sets of 10, plus 2 more individual units. Counting by groups also underlies the concept of multiplication. The product 3×7, for instance, means 3 sets of 7 objects each.

In a sense, modern society is all about modularity. Almost every aspect of life is compartmentalised. Specialisation abounds. Primitive societies of hunter-gatherers have relatively little specialisation. The women gather roots and berries while the males hunt game. However, if a family wants a hut to live in, they build it themselves. If they want water for cooking and washing, they fetch it. If they want clothes to wear, they make them.

In modern society, on the other hand, we pass all of the above tasks, as well as many others, on to specialists. The basics of survival include food, water, shelter, clothes, warmth. Obtaining these basics was a full-time occupation for early humans. Today, farmers grow our food, hydro authorities provide our water, and manufacturing industries provide us with the goods we need.

Instead of trying to provide necessities for ourselves, we gather them indirectly. We work at one thing—one job, one profession, or one speciality—not many. We use the money we earn to buy the basics we need from food, clothing and water supply specialists. This indirect approach allows society to provide more people with advanced goods and services.

We also use the divide and rule approach to organise our homes, our lives and our work. A typical primitive home was just a hut, a single space within which everything happened. Modern homes encapsulate different activities within different rooms: the bedroom is for sleeping, and the kitchen for storing food and preparing meals. All of this is so familiar that we do not normally recognise the essential truth—our customs and habits are designed to reduce complexity; to provide order in our lives.

In the workplace, computers provide many ways of applying this divide and rule principle. In the following sections, we will look at some of the technology that supports this approach.

Data modules

The idea of encapsulation, of carving things up into discrete units, is not confined to computer programs. We can carve up data in a similar way.

Suppose that a company wants to keep a database of its sales and clients. The data would consist of numerous *records*; each record providing the details of a particular sale.

Date	*Product*	*Price*	*Client*	*Phone*	*Address*
11 Jan	Widget	100	Nurk Inc.	666-999	11 Bush Ave
12 Jan	Gizmo	120	Klutz & Co	131-313	13 Luck Rd
12 Jan	Widget	100	Bloggs Ltd	123-456	12 High St
13 Jan	Widget	100	Klutz Coy.	131-323	13 Luck Rd
14 Jan	Gizmo	120	F. Nurk Inc.	666-999	11 Bushy Ave

Now if you look closely at this table, you will notice two things. First, there is a lot of repetition: details of products and clients are written down for every sale. Second, there are mistakes. If you have to enter values manually, then mistakes are inevitable. For instance, a name might be entered in a different way (e.g., Nurk versus F. Nurk), or there might be a typo in some contact detail (e.g., 131-313 versus 131-323). So here again, we see the same problems that plague programs: error and inefficiency.

The solution is to organise the data into modules. Separate out the different elements—sales, products and clients—and record them as three separate tables. First there is a table of products.

Product Code	*Product*	*Price*
P1	Widget	100
P2	Gizmo	120

Next, we have a list of all clients.

Client Code	*Client*	*Phone*	*Address*
C1	Nurk Inc.	666-999	11 Bush Ave
C2	Klutz & Co	131-313	13 Luck Rd
C3	Bloggs Ltd	123-456	12 High St

By referring to these tables, we can greatly simplify the final table of sales.

Date	*Product code*	*Client code*
11 Jan	P1	C1
12 Jan	P2	C2
12 Jan	P1	C3
13 Jan	P1	C2
14 Jan	P2	C1

This sort of arrangement is called a *relational database*. The idea is to save space and time, and avoid inconsistencies by setting up relationships between the tables. If the salesperson needs to type in the client's name, address and so on for a transaction, then they can simply select the name from a menu and the computer does the rest.

With such a small example, the reduction in size of the data table is not huge. In fact, in the above case the two extra tables actually make the new system slightly larger than the original. However, as the number of records grows, the space savings can be huge.

Encapsulation

A journey of a thousand miles must begin with a single step.
Lao Tzu, Chinese philosopher

As we have now seen, a modular approach to programs and data increases efficiency and reliability. You hide all the messy details in a module, a kind of software capsule. However, it can do much more than that. Encapsulation has some important properties. As we have seen, it simplifies design and reduces the possibility of error. It dampens the serendipity effect. It also removes the need to deal with low-level details.

In effect, the encapsulation of commands into modules creates a new universe within which to work. No matter how trivial, each 'module' encapsulates knowledge. For instance, the earliest macros simply bundled up sequences of commands. At the level of machine code, basic arithmetical operations involve a lot of shuffling of numbers between registers.

It used to be difficult to keep track of these registers, so mistakes were common. Hiding the details inside a module relieved programmers of the need to keep track of all the messy details. In other words, the earliest computer languages encapsulated the knowledge needed to get specific jobs done inside a particular machine. Encapsulating the basic details in modules meant that programmers no longer had to worry about them. Instead, they could write programs in terms of more complex ideas, such as arithmetic calculations.

This process of encapsulation is never ending. Once you have encapsulated simple elements in modules, those modules in turn become elements from which you can build larger modules. And so on.

The idea of encapsulation pervades all of computing. It also influences professional applications. For example, statisticians have developed libraries of routines to carry out the calculations that they perform frequently, such as taking averages, or fitting a curve to a set of data. In time, these libraries were turned into packages within which each operation could be identified by name. The packages also included the ability to write 'scripts', or programs that the package interprets and runs using the encapsulated operations. In some cases those scripts in turn created new languages. The same thing happened in all areas where encapsulation was used. For instance, the *Sequential Query Language* (SQL) became a standard way to write and manage database queries. Today there are high level computing languages that deal with information processing for many specialised fields.

The process of encapsulation has also influenced the organisation of information. Anyone who has worked in an office will be familiar with information hierarchies. Company records are traditionally stored in filing cabinets. The records department of a large company would typically run to several large filing cabinets, each one dealing with separate issues, such as staff, clients, inventory, and so on. Within each filing cabinet are drawers for different categories; each drawer contains folders and each folder may contain several pages of information. Computer manufacturers have deliberately echoed this hierarchy in their design of directories (sometimes called 'folders') for computer disk storage. It has also come to dominate the way in which the data are organised and stored.

In the earliest days of computers, there was just data. Suppose that you wanted to compile a list of great stories. Then in *runoff*, one of the earliest word processing systems, part of your list might look like this:

```
.p
.b Hemingway, Ernest
The Old Man and the Sea. USA. (1952).
.br
.b Shakespeare, William
Hamlet, Prince of Denmark. England. (1604).
.br
.b Shikibu, Murasaki
The Tale of Genji. Japan. (circa 1000).
```

Here the raw data contain what is known as 'markup'—commands to the processor that are inserted directly into the data. But these commands are concerned only with the fine detail of how the printed document is to be formatted (e.g., the command '.b' means 'bold face'). As we have seen, older databases would enter each author and title as a separate data record. More recent markup systems, such as the Standard Generalised Markup Language (SGML), and its Web-oriented descendent XML, echo that record structure. We use angle brackets (e.g., <author>) to indicate each unit of information.

```
<story>
   <author>Hemingway, Ernest</author>
   <title>The Old Man and the Sea</title>
   <country>USA</country>
   <date>1952</date>
</story>
<story>
   <author>Shakespeare, William </author>
   <title>Hamlet, Prince of Denmark</title>
   <country>England</country>
   <date>1604</date>
</story>
<story>
   <author>Shikibu, Murasaki</author>
```

```
        <title>The Tale of Genji</title>
        <country>Japan</country>
        <date>circa 1000</date>
    </story>
```

We can see that there is a hierarchy here. The element <story> contains other elements, such as <author> and <title>. In a real system, the hierarchy could contain many more layers, either upwards of downwards. For instance, the element <author> could be expanded to include each author's biographical details. Likewise, all of the <story> elements could be grouped to form a single element, say <bibliography>, that in turn, forms part of a larger data structure, such as the composition of an entire book. Here is a typical design.

```
<book>
    <frontmatter>
    <chapters>
    <endmatter>
        <bibliography>
        <index>
    </endmatter>
</book>
```

This design is very general. Each element can be expanded into further elements. For instance in a typical book, the front matter might include the title page, imprint page and table of contents and preface. Each of these elements can be expanded even further. The imprint page would include the publisher's name, address, date of publication and ISBN (book number).

As given here, the above design adequately describes the structure of most textbooks. If we omitted the details of those elements that form the end matter, then the design will describe just about any book at all.

The importance of the element approach is that it encapsulates each item in a systematic fashion. We can then refer to those elements in other contexts. For instance, below is a typical fragment of a template for creating a 'form letter'. The aim here would be to generate a series of letters by 'merging' names from a database with the text shown.

Dear <name>
Many happy returns for your birthday on <date>.

Here the elements <name> and <date> are treated as variables. When the document is processed they are replaced by values extracted from a database. Suppose our database contains the following entries.

```
<friend>
  <name>George</name>
  <birthday>June 27</birthday>
</friend>
<friend>
  <name>Mary</name>
  <birthday>August 23</birthday>
</friend>
```

After merging these details for George and Mary with the template, the final versions of the form letter would be:

Dear George
Many happy returns for your birthday on June 27.

Dear Mary
Many happy returns for your birthday on August 23.

From modules to objects

In the above examples, we saw that one application of encapsulation is to make it easier to use data. Now data are of no use without some way of displaying or interpreting them. For any given piece of data, there are typical methods that we use to deal with it. An example familiar to any computer user is the way that file names are provided with an 'extension'. The extension is used to define the sort of data the file contains, and we can assign programs to handle each type of data (the system usually makes default choices for us). So a file called FRED.TXT has the extension TXT. This suffix indicates that FRED is a text file, so the computer knows to open it with a text editor. Other extensions tell the system how to handle word processing documents, spread sheets, slide presentations, and so forth.

The obvious extension of this idea is to associate with each piece of information the methods that we want to use in processing it. In the example of a story shown earlier, the processing might be the format of the data for printing. So for the element <story>, the action might be to assign a new line, followed by processing the elements within it. Likewise, we might assign printing in bold face type to the element <author>. In the form letter example, we would need to link the database with the document template.

This idea of linking data and processes leads us to a new form of encapsulation, which is known as an *object*. The idea of object was introduced in 1962. Two scientists, Ole-Johan Dahl and Kristen Nygaard, working at the Norwegian Computing Centre in Oslo, were concerned about the *ad hoc* approaches being used in simulation. Their answer was to develop a programming language designed around the idea of objects. The aim was not only to simplify the development of simulation models, but also to provide descriptions of the elements. Called Simula, this language introduced an idea called a *class*. A class is like a cookie cutter for producing new objects. Suppose that you want to put together a model of lions and gazelles wandering around in the landscape. Each lion and each gazelle would be a separate object. But they are not completely different objects. Each lion shares similar kinds of properties (e.g., age and sex) and behaviour (e.g., sleeping, hunting, eating) with every other lion. So we can capture the idea of lion-ness (that is, the attributes and behaviour of lions) in the notion of a class of lions. And each particular lion is an instance of this class of objects.

In all the examples we have looked at, we can encapsulate data and processing to produce objects that each provide a particular kind of information. This approach, called *object-oriented modelling*, is fast taking over many areas of computing. For instance, the birthday form letter could be viewed as an object. It has two attributes, <name> and <date>, and its methods would be programs or scripts to merge and print the data with the letter. George and Mary are each instances of the class.

When we start looking around, we see that we can look at almost anything from an object-oriented viewpoint. For instance, a city is a class of geographical objects that have many common attributes (e.g., population, location). The object approach to computing has many advantages. The first is that it encapsulates everything we need to know about each entity—not only data, but also methods of handling the data. So, for instance, the city class of objects might include a method for drawing a city on a map. In this case, it would plot a city as a circle on a map, the size of the circle being determined by the population of the city.

Another advantage is that we can define relationships between different classes of objects. For instance, the class of cities that we described above really belongs to a more general class of objects that we might term 'inhabited places'. This class would not only include cities but also towns, military bases, research stations, mining camps, farms and any other kind of place that humans inhabit.

The city class inherits some of its attributes from the class of inhabited places. These basic properties would include its location, as well as (say) the current population. Other properties of the city class, such as contact details for the city offices, would be unique to it. The inhabited places actually form a hierarchy, with capitals inheriting properties from cities, cities inheriting attributes and methods from towns, and towns inheriting attributes from other inhabited places.

Classes can combine to form a new class. Drawings provide a good example. Let's suppose that we want to draw a stick figure of a person. We can build that figure, that object, out of simpler objects—head, body, arms and feet, each of which we can draw separately. But if we combine them, we get a new object—a stick figure. Many drawing programs allow us to do exactly this. We draw a collection of separate elements and then group them into the image that we want to save and reuse. So to draw a crowd, all we have to do is to paste copies of the stick figure on the page many times over.

These ideas are used routinely in graphics. For instance, the animation of stampeding wildebeests in the Disney movie *The Lion King* consisted of hundreds of separate wildebeest. Each wildebeest was animated by treating it as a

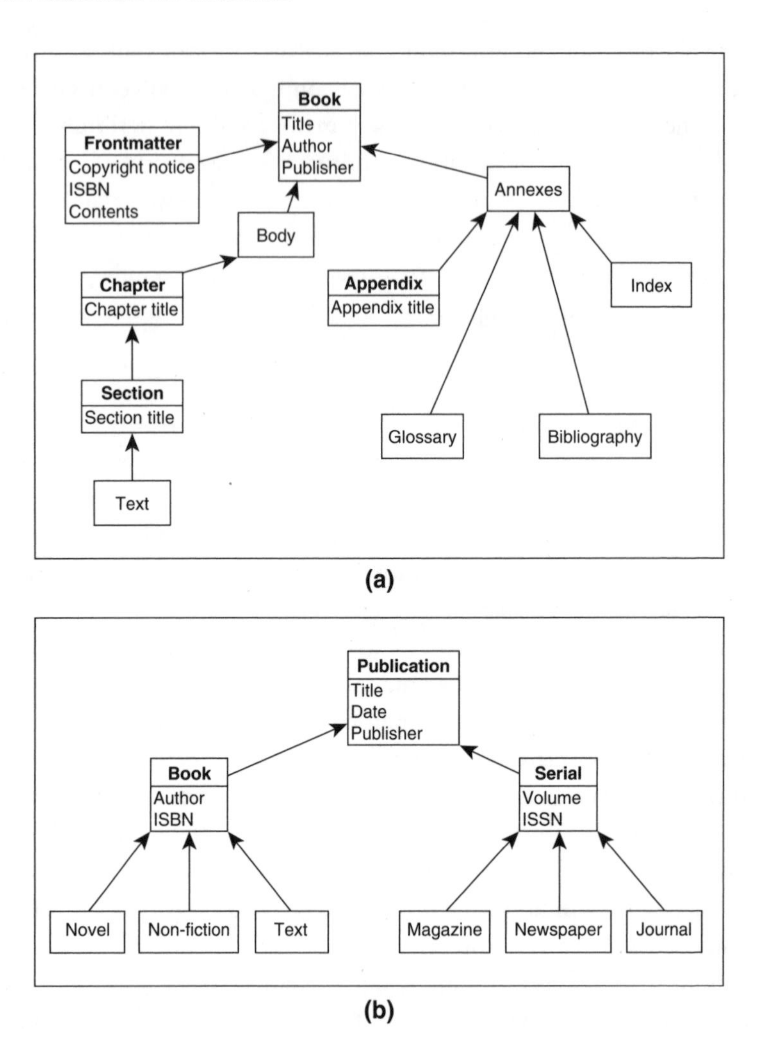

Figure 3.1 Hierarchies of classes associated with books
The name of each class of object is given at the top of each box. The words inside the boxes indicate attributes. (a) The structure of a book: many parts make up the whole. (b) A hierarchy of publishing objects. A novel is a special kind of book and a book is a special case of a publication. As well as its own attributes (e.g., Author), each class (e.g., Book) inherits attributes (e.g., Title) from more general classes (e.g., Publication).

separate object. All of the wildebeest objects behave in exactly the same way, but because they start in slightly different positions, and in different stages of motion, the overall effect is like the movement of a real herd of animals.

Building blocks

In the above discussion, I described the idea of an object as a way to simplify computing. However, the idea of objects is much more general than that—it is natural, it is even built into language. In fact, it can apply to almost everything. We use nouns to identify objects. We use adjectives and verbs to express their attributes and their actions.

Objects can be anything. The first kind of object that pops into your head is likely to be something solid, like a basketball or a tennis racquet, or a car. But, as we have seen, objects can also be abstract things such as lines of computer code or data structures. They can be drawings on a page, organisations, time periods, ideas or even feelings. Yes, even feelings such as love, or fear or hate, can be treated as objects. They may not have a physical size or shape, but people do assign attributes to them ('I feel blue') and they certainly do interact with one another.

The important thing about objects is that by bundling a number of objects together, we can create a new object, a new whole. This new whole is a universe unto itself, and its internal organisation is separate from everything on the outside. By grouping things in this way we are, in effect, saying that from the point of view of the outside world, internal composition and structure don't matter. We don't need to know about them. All we need know about is the whole object and its attributes, its behaviour.

This aspect of objects is perhaps the most important feature of modularity. It hides unnecessary details. We can build bigger objects out of it but it also simplifies things. We can see the forest, not the trees, let alone the cells that make up the trees. Human beings are composed of literally millions of cells, but we always deal with them as complete entities.

The nature of hierarchies

Hierarchies are familiar to everyone. Many human organisations, such as armies and corporations, are hierarchic in nature. Engineering systems, as well as complex software, usually consist of discrete modules with distinct functions. Large modules contain smaller ones, so forming a hierarchy of structure.

Hierarchies play a crucial role in complexity. The formation of modular structures is an essential mechanism for the emergence of order in many complex systems. For instance, the cells of a growing embryo begin to differentiate at a very early stage, forming modules that eventually grow into separate limbs and organs.

Hierarchies have an implied order, with a root node being the top (or bottom) of the tree. In systems composed of many elements, this order arises from the two ways in which hierarchies often form: lumping and splitting. Lumping involves objects coming together to form new objects (e.g., birds joining to form a flock). In splitting, a single object (say a system, class or organisation) breaks into two or more parts. Lumping and splitting are also associated with (respectively) the conceptual operations of generalisation and specialisation. For example, in architecture a building is a very general class of construction, one which includes much more specialised classes, such as a house.

In a tree structure, communication between elements is confined to pathways up and down the tree. That is, for data to travel from any node A to another node B, it must move up the tree until it reaches a node that lies above both A and B. It then passes down the tree to B. In a large system, a tree structure is an efficient way to ensure full connectivity.[5] The ability of hierarchies to connect nodes efficiently makes communication via the Internet feasible.

In management, hierarchies arise from the dual desires to simplify and to control complex organisations. They simplify by the divide and rule approach. A manager at any node need only be concerned with the nodes immediately above and below them. The model also restricts communication between arbitrary nodes in the tree. This enhances control, but can limit the passage of crucial information, thus inhibiting responsiveness, efficiency and innovation.

In engineering, hierarchies take the form of modules. As we have seen above, the advantage of modularisation is that it reduces complexity. Any large, complex system, such as an aircraft or a factory, may consist of thousands or even millions of individual parts. A common source of system failure is undesirable side effects of internal interactions. The possibility of problems grows exponentially with the number of parts, but can be virtually impossible to anticipate.

The solution is to organise large systems into discrete subsystems (modules) and to limit the potential for interactions between the subsystems. This modularity not only reduces the potential for unplanned interactions, but also simplifies system development and maintenance. For instance, in a system of many parts, there is always a risk that some part will fail. This is why engineers build redundancy into crucial systems. Modularity also makes it easier to trace faults. In some systems (e.g., computer hardware), entire modules can be replaced so that the system can continue to operate while the fault is traced and corrected. In computing, the idea of modules has led to *object-oriented programming*, in which each object encapsulates a particular function and can be reused in many different contexts.

An important effect of modularity is to reduce combinatorial complexity—useful in many kinds of problem solving. To take a simple example, if an urban transport system has (say) 100 stops, then it is essential that travellers can look up the cost of travel between any combination of stops. Now, if every trip has a different price (based on distance), then a complete table of fares needs to hold 5050 entries (assuming that the distance A to B is the same as that of B to A). This number makes any printed table unwieldy. On the other hand, if the stops are grouped into (say) five zones, with every stop within a zone considered equivalent, then the table would need at most 15 entries, which would make it compact and easy to scan.

What is true of computers is also true of cities. To support huge concentrations of people, cities have to provide a wide range of services: housing, transport, distribution of food and other commodities, water, sewage, waste disposal, power and communications. On top of these basic amenities, there are social infrastructures such as education, hospitals, emergency services and shopping centres. The interactions of so many systems, combined with external factors such as rapid growth, technological development and social change underlie many problems in modern society.

Although hierarchies reduce complexity, they can introduce brittleness into a system. That is, removing a single node, or cutting a single connection, breaks the system into two separate parts. Every node below the break becomes cut off. This brittleness occurs because hierarchies are minimally connected. There is no redundancy in the connections. For instance, the Internet is organised as a hierarchy of domains. If a domain name server fails, then every computer in that domain is cut off from the Internet. Centralised services, such as city power supplies, also suffer from this problem. This problem of brittleness is intimately associated with the platypus effect, a phenomenon associated with hierarchies. In the next chapter, we will look at the platypus effect and its many implications.

chapter 4

THE PLATYPUS EFFECT

If order appeals to the intellect, then disorder appeals to the imagination.

Paul Claudel[1]

In 1798, scientists at the British Museum of Natural History were dumbfounded when they began to examine one of the stuffed specimens that had recently been sent to them from the colony of New South Wales. The animal was unlike anything they had ever seen before. It defied all the known rules of science. The creature was covered in brown fur like an otter, and suckled its young, so it was obviously a mammal. And yet it laid eggs, just like a reptile. It also had webbed feet and a bill, just like a duck. To top things off, it had a tail like a beaver.

The sceptical scientists could not bring themselves to believe that the thing was real. Such an animal could not possibly exist, they told each other. They concluded that it must be a fake foisted on them by troublesome colonists. Convinced that it had been sewn together from parts of different animals, the suspicious scientists took a scalpel to the specimen to try to locate the stitches. Many years passed before scientists were forced to admit that the platypus was genuine. It took a living specimen to convince them.

The platypus really is an oddball of an animal. In 1798, it did not fit into any of the known taxonomic categories. Much later, the platypus was found to be a missing link between mammals and reptiles and today we know that it belongs to an ancient group, the monotremes, which split off from other mammals millions of years ago. Only one other living monotreme species, the echidna, or spiny anteater, still survives.

The platypus lends its name to a phenomenon, the *platypus effect*, that occurs whenever things do not quite fit into the plans that we have laid down. It is actually a variation of the serendipity effect, but instead of surprise discoveries, it leads to things going wrong. As we shall see in this chapter, it is a widespread phenomenon, especially when we attempt to organise information.

Although the name *platypus effect* is associated with taxonomy, the term was coined not by biologists, but by software engineers. They used it to refer to a problem that they encountered when trying to design information systems. The usual approach is to carve up the system you are trying to describe into distinct modules, each of which deal with a major aspect of the problem at hand. (We saw this divide and rule approach in the previous chapter.) For instance, in describing a business, one module might deal with the products, another with staff and suppliers, and another with customers. The platypus effect is seen when, having carved up the system, engineers encounter something that does not fit neatly into any of the modules that they have created. The above list of business matters, for instance, makes no mention of assets and other company holdings. To accommodate new areas such as these, an engineer would either have to generalise the existing modules, or else add new ones. The problem usually arises when existing parts of a system turn out to interact in unexpected ways.

We see the platypus effect at work throughout science. Traditionally scientists have divided knowledge into different fields of study. Mention physics, chemistry, biology, or mathematics, and everyone knows what you are talking about. The names are enshrined in the titles of university departments the world over. But herein lies a problem. Nature does not divide itself up in such neat ways. Many, perhaps most, of the most active areas of modern science involve phenomena that lie between traditional areas of knowledge. The majority of the resulting interdisciplinary fields of study have double-barrelled names, such as evolutionary computing, environmental health, geographic information, quantum electronics, socio-economics, or molecular medicine.

Some fields of study today are interdisciplinary by their very nature. One reason for this is the serendipity effect. There are lots of potential interfaces between disciplines, so it is only to be expected that some of them will give rise to important discoveries. For example, in later chapters we will see that both biotechnology and environmental management depend heavily on information technology.

For many years, and in many institutions, people have struggled with research that crossed disciplines. The field of complexity, for example, was long ignored because it cut across so many disciplines. It emerged as a coherent discipline only when computer models started to become common in different fields.

Getting in a muddle

The best laid schemes o' mice an' men
Gang aft a-gley.

Robert Burns[2]

'It's in here somewhere,' says Mary Sue as she rummages through her handbag. Her boyfriend waits patiently while she searches for her lost car key. But as the seconds turn into minutes, he suspects the worst.

> 'Are you sure you didn't lock it in the car?'
> 'No, it's definitely here somewhere. I remember putting it away.'
> 'Well why don't I go and look in the car, just in case?'
> 'No, wait a minute. Aha! Here it is!' Triumphantly Mary Sue holds up the lost key.

How many times have you experienced something like the above scene? Or what about the following? You are rushing out the door, late for work, when suddenly you realise that you don't have your reading glasses. (Or perhaps it was your pen, your wallet, your umbrella, or some other essential item.) You rush back inside, but the glasses are nowhere to be found. You look in all the usual places. Nothing. You look in some unusual places. Still nothing. Time is racing. You are getting desperate. Then you remember that you fell asleep on the couch last night. So you check the couch. Lo and behold, there they are, hidden under a pillow.

These tales, and many more like them, get played out a thousand times each day in households all over the world. Why do they happen? Along with other everyday problems, they are symptoms of complexity. A woman's handbag may sometimes contain hundreds of items. So if she needs to find one particular object, she may need to check many other items before she finds it. But what is worse, the items interact with each other. Just as glasses hid under a pillow, the key may be lurking under her compact. So if she just shuffles the compact, but does not lift it out, then the keys may remain hidden from view.

A solution to the above problems is modularity, that is, breaking the large problem down into simpler, more manageable ones. Handbag manufacturers are well aware of this and usually provide many pockets and compartments to help you organise your belongings. Put your keys in the key pocket and you can find them in a trice. The pockets are a form of modularity.

As the number of items that we have to sort grows larger, the need for modularity increases rapidly. Without compartments, or some other form of hierarchical organisation, we are lost. Entering Dr X's office is like a field trip into a knowledge jungle. There are books and papers piled up everywhere. No hint of a desktop is to be seen. It is hidden under layer upon layer of papers, books, memoranda and notes. Ask Dr X for a journal article from last year and he can find it in seconds, but no-one else could divine order in those mountains of paper.

No business can survive without a viable filing system. As we saw in Chapter 3, files are highly modular. You place the invoice from company Y in the file, and the file in a drawer and the drawer in a cabinet. This organisation makes it possible for you to find any item quickly by working your way down through the hierarchy.

Nevertheless, the platypus effect is always lurking, waiting to strike. To understand it, you don't need to go any further than your own home or office. Take your incoming mail. Some of it you can junk in the cylindrical file immediately. But where do you store the rest? Let's say you start by dividing them into two categories: *financial* (bills to pay) and *personal* (letters from friends and family). These categories account for most of the mail, but then along comes the summer sales catalogue. You deal with it by creating a new pigeonhole for *commercial information*. Then a letter arrives inviting you to list your company in a local community services directory. Again, you create a new category, this time for *publicity*. Next, an invitation arrives asking for your participation in a local community event. This letter is not personal; it relates to your work. Nor is it really publicity. You create a new category called *community*. And so it goes on. Each new category you invent accounts for more and more of the incoming mail but, always, there will be a fraction of incoming mail that doesn't quite fit your current categories. This is the platypus effect in action.

Much the same thing happens when we try to clean up the house. Most household items have a 'home'. Food goes in the kitchen. Clothes go in the bedroom. Tools go in the garden shed. But what do you do about all the odd things? What about your holiday souvenirs? What about that dreadful wall hanging that your uncle gave you for Christmas? No matter how hard you try, you can never be completely organised. Anyone who has ever moved house knows that moving involves organising your belongings. Every time you move, items go missing. They are somewhere in the new house, but you simply cannot find them. Sometimes, items disappear for years on end and mysteriously reappear the next time you move!

Modularity reduces complexity. Leave items lying around anywhere and you are bound to have pillows hiding them. But put them away where they belong and the complexity of the system plummets almost to zero. We tend to find the greatest level of modularity in systems with the greatest potential for complexity. Libraries, for instance, contain thousands, sometimes millions, of books. Without an indexing system, no-one would ever find anything. In a library, we find modularity taken to the extreme. The Dewey system, for instance, provides numbers for a thousand categories and subcategories. But fields can be further refined by adding decimals at the end of the number. You can add the author's initials too, for good measure.

One of the dangers of modularity is that it is brittle (Chapter 3). One of a librarian's worst nightmares is for a book to be placed on the wrong shelf. Put a book in the wrong place and you might never find it again. Librarians are very careful to replace books correctly. Readers, however, are the bane of their existence. Readers are likely to browse through a book and put it back on a shelf, any shelf. This is why most libraries ask readers not to return books to the shelves. Despite the warnings, books simply disappear within the system. To address the problem, small libraries close their doors every so often and carry out a complete inventory of the shelves. This is not practical for larger libraries. They avoid the problem by keeping the readers away from the stacks altogether. If you want to read a book, then you submit the call number to a librarian who collects it for you and returns it to the shelves when you are finished.

Public servants face a similar problem when storing correspondence. Although they have a catalogue system for grouping files, the headings under which records are stored may not be transparent to outsiders. They get around this problem by duplication. For instance, they store a second copy of outgoing letters by their date and include a cross-reference to the thematic records. That way, they have at least two different routes by which to find an item.

This reshelving problem also explains why Mary Sue lost her keys at the beginning of this section. Being in a hurry, she threw them into her purse without thinking. Had she taken the time to put them back in the pocket where they belonged, she would have found them instantly.

This is a problem that we all face. If everything is perfectly sorted, then nothing gets lost. But keeping things in order takes a lot of work. Any parent can testify to that. A certain degree of organisation is essential to any activity: businesses need orderly records so that they can keep track of income and expenditure; libraries need to record not only the position of every book on the shelves, but also every time that a reader borrows a book; stores and warehouses need to keep thorough records of the movements of inventory. But if you try to keep everything in perfect order, then you are likely to find that you have no time left over for anything else. So we are forced to make a compromise between order and chaos.

What animal is that?

Taxonomic classification is a practical example of knowledge hierarchies. Biologists often use taxonomic keys to help them identify plants and animals. A traditional key consists of a series of 'yes' or 'no' questions. Each answer narrows down the range of possibilities until at last you reach the name of the animal you are trying to identify. Look at the following simple example to get the idea.

1. Does it have sharp teeth?
   YES Go to 2.
   NO Go to 5.
2. Does it swim?
   YES Go to 3.
   NO Go to 4.
3. Does it live in the ocean?
   YES It's a shark.
   NO It's a crocodile.
4. Does it have a mane?
   YES It's a lion.
   NO It's a bear.
5. Does it eat grass?
   YES It's a horse.
   NO Got to 6.
6. Does it have a long neck?
   YES It's a giraffe.
   NO It's a koala.

This sort of key is very easy to use. Just answer the questions and the solution pops out. Admittedly, the above key is extremely simplistic, but it does imitate the form that many traditional keys take. Real keys usually start by narrowing down the standard taxonomic details: class, order, family, genus, species.

The above example also serves to highlight some of the drawbacks of this approach to classification. First if all, it suffers from the platypus effect. The number of animals in our key is obviously extremely limited. Suppose that the animal you are looking at is a blue whale. Then your answers would be as follows:

1. No, it does not have sharp teeth.
5. No, it does not eat grass.
6. No it does not have a long neck.

With these answers, the key would tell you that the animal was a koala! But the blue whale also possesses features listed under other categories; for instance, it lives in the ocean. So here we have an animal, like the platypus, that defies the given classification framework. The limitations of our example are obvious, but real information systems can fail for the same reason.

Another drawback is that the key is very 'brittle'. It is easy to go wrong. Just like lost keys in a handbag, animals can get 'misplaced'. If you make a single mistake, then your classification is inevitably wrong. For instance, what does the term 'sharp teeth' mean? If you look into a horse's mouth, it certainly has sharp teeth. The point is that they are used for cutting grass, not flesh. So this one misunderstanding would cause you to mistake a horse for a bear!

Given this brittleness, biologists have experimented with alternative, more robust, methods of classification, especially in automated keys. One method is to make up a profile of each species against all the characters. You can then rank the possibilities according to how well they fit the profile.

The question of classification takes on greater urgency when we realise that doctors use similar methods to diagnose diseases. Here the problem becomes much more acute, not only because human lives are often at risk, but also because the platypus effect is more pronounced. There is a multitude of known diseases and many share at least some signs and symptoms. To make matters worse, the symptoms associated with a single disease can vary considerably from one patient to another. So it is not surprising, but far from reassuring, to learn that misdiagnosis is all too common. This is one reason why doctors frequently send samples off for analysis to confirm their suspicions. Likewise, the profile approach has attracted interest among scientists trying to automate medical diagnosis. One of its advantages is that it does not exclude possible diseases on the grounds of obscure symptoms.

Are birds dinosaurs?

One reason why the platypus effect occurs in taxonomy is because the classification hierarchy does not always reflect what really happens in nature. To understand this, look at the case of the birds. There has been a long, heated debate over the origins of birds. Are they descended from dinosaurs?

Taxonomic categories do not reflect evolution very well. Consider four well-known groups of large animals: mammals, reptiles, dinosaurs and birds. Most of the confusion arises because groups are named according to the animals that survive today. Birds, reptiles and mammals were classified as distinct groups long before they were known as fossils. When dinosaurs were discovered, they were classed as reptiles because they shared some obvious characteristics with reptiles, and they evolved from reptiles anyway. Mammals, however, which also evolved from reptiles, are classed as a separate group because they still exist. This inconsistency has coloured the way people think about dinosaurs. Until the discovery of *Deinonychus* and other highly active dinosaurs, it was generally assumed that dinosaurs were sluggish and cold-blooded, just like modern reptiles. This preconception also precluded them from being related to such obviously active animals as birds.

Classifying animal groups based on modern survivors has been the source of considerable confusion. If the platypus and echnida existed only as fossils, biologists would probably classify them as mammal-like reptiles. This is, in part, due to what Stephen Jay Gould calls the 'shoehorn' effect. That is, people have existing categories, so they try to fit new things into those categories.

We can see this effect at work in another taxonomic example. In his book, *Wonderful Life*, Gould describes the confusion that followed the discovery of a rich bed of early Cambrian fossils in the Burgess Shales of western Canada.[3] Charles Doolittle Walcott discovered the Shales in 1909 and, over the next 15 years, collected more than 65 000 specimens from the site. When he came to describe his findings, Walcott was confronted by thousands of weird looking fossils from 500 million years ago. He assumed that the specimens were bizarre early members of known taxonomic groups. It was not until some 50 years later that scientists re-examined the fossils and realised just how bizarre they were. In 1969, Cambridge palaeontologist Harry Whittington, together with his students Conway Morris and Derek Briggs, began re-examining the fossils. Much to their surprise, they found that many of the specimens did not fit within existing taxonomic categories. They had to assign 21 species to entirely new phyla. With just 15 previously known phyla represented, the fossils suggested that more than half the animal phyla that ever existed have since been lost. To put this in perspective, all of the familiar, large animals that we know today—all the mammals, the birds, the reptiles, fishes, and so on—were represented by just two species from a single phylum, the chordates.

Here we have another example akin to our original story of the platypus. Confronted by something completely outside their experience, scientists at first failed to recognise genuine novelty when they saw it. It is perhaps no accident that Walcott's traditional interpretation was accepted without question for over 50 years. There is even a suspicion that had he published a modern interpretation of the Shales, his peers would have rejected his story. The truth was simply too big a mental leap to make at the time.

The muddled platypus

In 1999, there was a humorous photo doing the rounds of the Internet. Labelled as the winner for the 'Not My Job' award, it showed a squashed porcupine lying in the centre of the road—hardly the most savoury of topics. The funny thing was that road workers had painted the centre line for the road right over the top of the corpse. Their job was to paint lines, not to dispose of road kill!

The platypus effect arises when we try to divide a complex system into modules. Some of the best examples are found in large organisations, where the modules may be corporate divisions or government departments. The military have the most formalised modular structures. An army is divided into corps, divisions, brigades, battalions, companies and platoons. An officer of the appropriate rank manages each unit in the hierarchy.

The hierarchy model also describes the divide and conquer approach to problem solving, which we met in Chapter 3. That is, you break a large, complicated problem down into progressively smaller, simpler ones until you can solve them. In science, this method is also known as the reductionist approach.

The modular approach to problem solving is highly successful. It reduces the incidence of unexpected interactions enormously. Take any large problem, such as governing a country, and you can simplify it by dividing it up into smaller units, such as states, provinces, or districts. The smaller units are not restricted to smaller versions of the same thing. A company, for instance will divide its operations up according to function, creating separate divisions such as purchasing, supply, maintenance, production, finance and advertising. However, carving up the work in this way sometimes creates problems. Problems such as the dead porcupine almost always arise when work is carved up into different areas of responsibility. Inevitably, something will crop up that falls between two departments. This can sometimes have tragic results, as in the case of an elderly couple, though eligible for free government housing, being evicted from government housing for failing to pay their rent. In business, it can mean the difference between success and bankruptcy. How often are opportunities lost because they fall in the gap between different arms of an organisation?

One of the great triumphs of the Sydney 2000 Olympic Games was how smoothly the organisation ran. Given the enormous scale of the event, and the need to coordinate the operation of countless different organisations, the opportunities for foul-ups were huge. Fortunately, the organisers took the smart course of employing a team of devil's advocates, whose role was to try to find holes in the planning. They asked everyone the awkward questions and they posed tricky what-if scenarios.

To sum up, carving things up into modules simplifies things, but at the same time there is also a risk that the platypus effect will create unexpected problems.

Finding your way

You can find a good example of modularisation reducing complexity in road networks. Most older cities around the world predate motor traffic by at least a hundred years. With no urban planning until relatively recent times, facilities developed almost at random. In particular, many road networks are idiosyncratic. For example, taxis travelling from downtown Sydney to Kingsford Smith airport have to negotiate a weird sequence of back streets, highways and intersections.

Similar problems plague other highway networks, particularly the main arteries leading into many cities. In Cape Town, South Africa, for instance, a study commissioned by the Cape Metropolitan Council (CMC) in 2000 concluded that the city would soon face urban traffic chaos. Over 670 000 vehicles regularly travel in greater Cape Town, a number which had increased by 80 per cent over the previous two decades. A similar, unchecked increase would see gridlock in the metropolitan area, in which over 80 per cent of all jobs were located.

In many Asian cities, massive traffic congestion has almost brought the city to a halt. On a visit to Seoul, South Korea some years ago, my hosts declined to take me across town for some sightseeing because driving there would have taken hours each way! The subway was faster. During the early 1990s, Bangkok in Thailand had one of the world's worst problems with traffic congestion. In a study that outlined proposals for rectifying the problem, Dr Kanchit Pianuan, an engineer from Chulalongkorn University, summarised the problem as follows:

> Bangkok's traffic situation is desperate. And yet the problem has never been properly analysed. The current chaos is an inescapable result of an endless hodgepodge of half-measures, wrong measures, and no measures. Expanding existing roads, for example, merely causes further congestion due to messy, obstacle-course construction that hurts at least as much as it helps. It also encourages people to put more cars on the roads.
>
> Bangkok's average traffic speed rates are much lower than those of other cities. As traffic gets worse, the rates drop even more. The plethora of business centres along the Ratchadapisek Road, for example, has slowed traffic there to a maddening crawl, just 8 to 9 kms. an hour in normal traffic situations, and a barely moving 2 to 3 kms. an hour in rush hours. In crowded residential areas, traffic flow is now only 10 to 15 kms. an hour. The city centre, without a single restricted zone, of course has the most severe traffic. People do not drive there any more unless they must.

> Most data on Bangkok's traffic problem is compiled based on the personal preferences of those involved. This is grossly inadequate for solving a problem of this magnitude and complexity. Systematic and coordinated planning is needed.[4]

In contrast, planned cities can be a dream for drivers. A good example is the Australian capital, Canberra, a planned city with a strictly modular design. The city is spread over nearly a thousand square kilometres and is organised around four major town centres. Each of these centres is further divided into suburbs, and each suburb conforms to a standard plan. A typical suburb has a shopping centre and a single major street running through it. Other streets and avenues branch off this main thoroughfare. The suburbs are joined by major roads and a system of freeways links all the town centres.

This hierarchical system of roads makes driving from A to B in Canberra very simple. Just exit the suburb you are in, take a freeway to the area you want, then enter the next suburb you want. Wherever you start from, you can get to your destination in less than 30 minutes. In normal conditions, traffic jams are rare. The city is so decentralised that on my first visit to the city centre, I thought that I had landed in a suburban shopping centre by mistake. As with any hierarchy, however, Canberra's road system is brittle. Transport between major centres relies heavily on freeways, so any disruption such as an accident can cause massive bank-ups of traffic. When my wife was in labour with our first daughter, it was a wet morning and the only highway to the hospital became jammed. By the time we found a way around, the trip had taken twice as long as normal!

The revenge effect

In his 1996 book *Why Things Bite Back*, Edward Tenner describes an ironic aspect of technology.[5] New technologies are introduced to solve problems. In many instances, however, the introduction of a new technology only ends up making the problem worse. Tenner called this phenomenon the *revenge effect*. Early sunglasses, for instance were meant to protect people's eyes from bright sunlight. What they did, however, was cut down the amount of visible light. This made the pupils dilate, letting in more ultraviolet light, which does the real damage. Likewise, the introduction of the cane toad to Australia to control insect pests unleashed an even greater pest than the one it removed.[6]

Computers produce as many revenge effects as any other technology. One slogan that manufacturers have used to help sell their machines is the idea of the paperless office. The idea was that on-screen display, email and file sharing would make it unnecessary to shuffle paper around. Paradoxically, the introduction of computers has had exactly the opposite effect. Most offices now go through more paper in a year than ever before.

Even more prevalent than Tenner's revenge effect are the side effects that accompany technology. In this respect, the revenge effect is closely related to the platypus effect.

One peril of the ever-present drive for corporate efficiency is the increasing brittleness that goes with it. For instance, in his book *Faster*, James Gleick describes the relentless drive of airlines to achieve ever greater efficiency.[7] In doing so, they are shaving away all of the redundancy that allows them to cope with the unexpected. If a single aeroplane is found to have a fault, there will probably be no way to replace it quickly. Flights and schedules can be thrown into chaos. The following little horror story shows what can happen.

In 1994, a complicated itinerary required me to fly from Auckland, New Zealand, to Sao Paulo, Brazil via Los Angeles, USA. As my plane touched down in Auckland, everything seemed on time and in order. But then there was an announcement that the plane had struck a bird while landing and had to be checked for possible damage. This, we were told, would involve a slight delay. The 'slight delay' turned out to be three-and-a-half hours. The only reason the flight eventually took off was that there were not enough beds in town to cope with the hundreds of passengers on board. It was the only time I've ever known an airline to serve a meal, and show a movie, while still on the ground! We did take off, but now the problems snowballed. The flight crew had been on duty too long, so we had to divert to Honolulu to allow them to be relieved. But this detour involved a longer flight, which meant that we also had to refuel. By the time we arrived in Los Angeles, we were over nine hours late. Although I had allowed a generous margin of six hours to make my connection, it had long gone. So we queued up, over 200 of us, for another three hours while the ground staff tried to reschedule our flights. My case was so difficult that at one point they were going to send me hopping from city to city all over the US trying to catch an appropriate flight. I arrived in Brazil some 30 hours late. After a taxi ride of over 100 kilometres, I finally reached my destination just in time to attend the last afternoon of a three-day meeting.

Breakdowns

At 5:25 p.m. on 9 November 1965, a series of power failures blacked out vast areas of the northeastern United States.[8] Cascading failures struck the overworked power grid without warning. The blackout extended from Buffalo in the west to New Hampshire in the east and from as far south as New York City to Ontario, Canada in the north. Nearly a week later, the cause was traced to a single faulty relay switch. This one small failure had led to an escalating series of failures. Like falling dominoes, one after another system became overloaded. In New York City, the power generators shut themselves down to prevent their turbines being overloaded by surges from the grid.

The blackout left about 30 million people without power until 7 a.m. the next day. Surprisingly, there was relatively little looting and other crime. According to some reports (possibly based on urban myths), the event was marked by a spike in the birthrate nine months later. Twelve years later, in 1977, a second extended blackout in New York did lead to widespread looting and mayhem.

A single, unexpected event can disrupt major services—water, gas or electricity—for an entire city or even an entire state. Such cases arise from a combination of two things, serendipity and fragility. Central services are, by their very nature, extremely fragile. If one service goes out of action, for whatever reason, then the entire system fails. The power supply of every city gets knocked out at some time or other. And because power is centralised, a single outage can affect millions of people.

The New York blackouts are not unique, of course. What was unusual about the 1965 blackout was the vast area affected. Such blackouts not only show the platypus effect at work, but also highlight the brittleness of complex, centralised facilities.

chapter 5

FROM THE NET TO THE GRID

Computers make it easier to do a lot of things, but most of the things they make it easier to do don't need to be done.
Andy Rooney, CBS News correspondent

On big problems

What is the largest number you can think of? A million? A billion? School students sometimes play this as a game. In 1940, Edward Kasner and James Newman gave the name '*googol*' to an astonishingly large number that consisted of 1 followed by 100 zeroes.[1] But the problem with any large number is that there is always a bigger one. They gave the name *googolplex* to 1 followed by a googol of zeroes. Likewise, a *googolplexplex* would be a 1 followed by a googolplex of zeroes.

In everyday life, most people have scant need to know about such enormous quantities. And yet, when we have complex problems to solve, huge numbers can arise very easily. As we saw in Chapter 2, combining different things leads to complexity, and hence to serendipity, as well as other unexpected effects.[2] The number of possible combinations and arrangements of things increases at an exponential rate and can easily become astronomical.

Some problems are so large, so hard, that people have simply tried to avoid them. Although the divide and rule principle (see Chapter 3) often helps, solving many kinds of problems reduces to a matter of trial and error. You take potential solutions, one after another, and test whether they really do solve your problem.

The *Travelling Salesman Problem*, or TSP for short, is perhaps the best known example of a hard problem. The challenge is to find the shortest path that a salesman can take in visiting a number of cities. He must return to his point of origin at the end, but he visits no other city twice.

This problem is hard to solve because it involves looking at every possible way in which the sequence of cities can be ordered. If you wanted to find the shortest distance between any two towns, then you would need to compare the distance between every possible combination. For N cities, there are $\frac{N(N-1)}{2}$ combinations you need to check. This number is a polynomial, so we say the problem is *P complete*: it can be completed in polynomial time. The TSP is termed *NP hard*. This means that the search for a solution cannot be completed in a time that increases in proportion to some polynomial of the number of cities.

To solve the TSP by an exhaustive search is normally not feasible. For just 30 cities, using all the world's computers, the task would take longer than the age of the universe! However, we can get very good answers in reasonable time. The graph (Figure 5.1) shows a typical example of a completely random trial and error search for solutions to the TSP. Notice that every so often, the random trials throw up a solution that is better than any considered before. However, the curve quickly flattens out because those improvements become increasingly rare.

If you have big problems to solve, then you require a lot of computer power. For this reason, scientists and engineers have always sought to build bigger and faster machines. By the 1970s, there was a lot of prestige to be gained by saying that you had the world's fastest computer.[3] Some well-known problems, such as the TSP, become *benchmarks*. They are used to evaluate and compare the performance of different methods and different machines.[4]

Many hands make light work

Which would you rather have to pull your wagon—a thousand chickens or one good, strong ox? In the early 1990s, the proponents of high

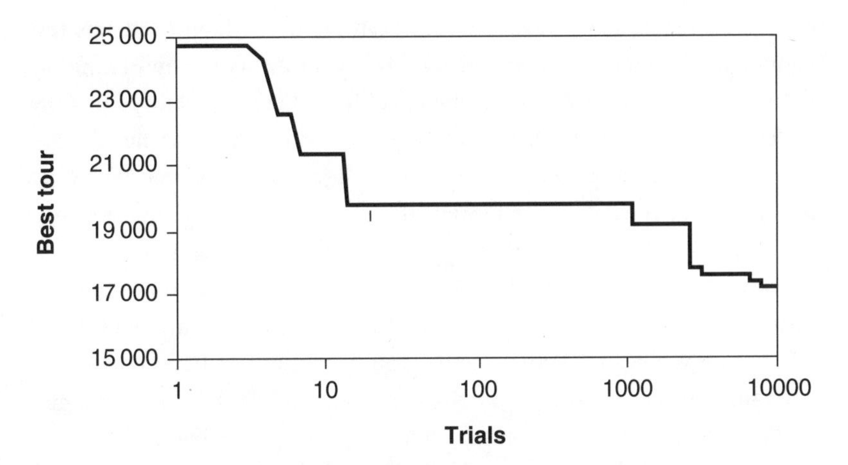

Figure 5.1 A random trial and error search for solutions to the Travelling Salesman Problem when there are 100 cities

The bottom axis indicates the number of trials competed. The vertical axis shows the best time achieved. Each trial consists of listing the cities in random order and calculating the length of the resulting tour.

performance supercomputers posed this rhetorical question in an attempt to deter people from buying massively parallel machines. Big problems need big machines. Surely? The thought of harnessing a thousand chickens to do useful work is so absurd, that at first sight the choice seems obvious. But is it?

Imagine a busy shop. People choose the goods they wish to buy, then line up at the counter to pay. If there is only one shop assistant on duty, then each customer can face a lengthy wait before the cost is totalled up, payment sorted out and the goods packed. If the shop is really busy, then customers arrive at the checkout faster than the cashier can process their purchases. As a result, the queue grows longer and longer and customers become impatient. So how do you speed up the queue? As everyone knows, you just add more checkouts. A large supermarket during a busy period can have 30 or more checkouts active at once.

Parallel computers adopt the same approach to problem solving. You cannot speed up the processor inside a machine, but you can add more processors. Perhaps the most extreme example of this approach was the Connection Machine, developed by Danny Hillis' company Thinking Machines Corporation (TMC). The CM2 model contained up to 65 536 processors, all packed inside a black box the size of a large cabinet. So the machine was like thousands of computers all jammed together. By 1989, the CM2 had proved the potential of parallel computing by achieving processing speeds in excess of most supercomputers at the time.

Unfortunately for TMC, as well as many other supercomputer manufacturers, the demand for these huge machines is not great. Only large organisations can afford them, and desktop computers have become so powerful that they are now more than adequate to handle most problems. TMC was forced to file for insolvency on 17 August 1994 and did not survive. Several other specialist high performance computer companies have either followed suit or been taken over by companies with a wider range of computer products.

In 1998, Arthur Marcel described the situation as:

> In the 1980s, there was a trend for big, totally integrated information systems. Many of these either failed or never got going. It is now recognised that a network of smaller, localised sub-systems, each handling their own particular area in their own particular way, is superior. Not only are they easier to implement and maintain, but each sub-system evolves to best meet the needs of its particular service sector.[5]

Despite the commercial problems of individual companies, parallel computing lives on. It did not take long for people to realise that if you can make a large, parallel computer out of many simple components, why do those components have to sit in the same box? You could achieve much the same thing by linking together a number of separate computers so that they can share data efficiently. The Beowulf Project, for instance, has developed designs and software for turning networks of ordinary desktop computers, running under the freeware operating system *Linux*, into a parallel-computing cluster.[6] Essentially, we are seeing the beginnings of do-it-yourself super-computing. One day, I entered our research centre's computing lab to find that we suddenly had a 'supercomputer' in a back room. David Newth, a research student in the school, had linked several old, discarded personal computers around the office into a Beowulf cluster.

Many computer centres now look somewhat empty as banks of PCs lining the walls replace the huge cabinets which housed the supercomputers. Big problems do not necessarily need big computers to solve them. As we have seen, we can often achieve the same result, and perhaps even do better, by distributing the workload among many small machines. By 2002, the TOP500 index listed 80 clusters in the top 500 supercomputers.[7] At least seven of these clusters consisted of PCs. The fastest computer was the NEC corporation's Earth Simulator, a special-purpose machine consisting of 5120 processors.

But if a thousand chickens are better than one big ox, then perhaps a million ants are even better? If lots of small processors can provide more computing power than one large processor, then do they even need to be tightly linked together into a cluster? Not all problems need the different processors to talk to each other. For instance, if you want to check the properties of a million different car designs, then checking each design is a separate problem that can be carried out independently of checks on all the others.

Problems that can be farmed out this way can be carried out perfectly well by many small computers, each working on their own small piece of the puzzle. With the Internet's ability to link computers all over the world, many people soon began looking at ways to achieve what we might call *communal computing*.

Communal computing

In 1967, radio astronomers at the Cavendish Laboratories in Cambridge made an extraordinary discovery. They detected radio signals containing short, regular pulses. At first they suspected interference from some man-made source but, when it was confirmed that the signals really did come from outer-space, the mystery intensified. At the time, there was no natural phenomenon known that produced precisely regular beats of the kind they were receiving. The news caused a sensation. Newspaper reports speculated that at last humans had found evidence for life in outer-space.

Unfortunately, it was not to be. Subsequent research found that the signals emanated from an entirely new kind of celestial object that came to be known as a pulsar.

The flurry of excitement over the discovery of pulsars sparked public interest in searching for alien signals. Already in 1960, radio astronomer Frank Drake had begun a systematic search of the sky (Project Ozma) to look for radio signals from alien civilisations. The *Search for Extraterrestrial Intelligence* (SETI) is a scientific movement that aims to find intelligent life elsewhere in the universe by looking for radio signals from outer-space. The idea that inter-stellar civilisations try to contact each other by radio has featured prominently in science fiction over the years.[8] If this is so, then intelligent life out there is probably trying to send us messages already. The challenge for SETI projects is to detect potential signals. In practice, this means looking for abnormally high numbers of spikes at particular radio frequencies.

Funding is a major problem for SETI. Several SETI projects run by NASA were closed in 1993 because of increasing budget constraints. Much of the raw data from SETI comes from the Arecibo radio telescope in Puerto Rico. Because the telescope is fixed, it cannot be aimed. So there are periods when other astronomers cannot use it. However the biggest problem is that looking for an alien radio signal among millions of stars is like looking for a needle in the proverbial haystack. Processing radio signals from space requires an enormous amount of computing power. SETI projects have used supercomputers to process the data, but supercomputer time is expensive, and limited.

Then SETI scientists hit on a novel idea. Owners of personal computers typically use only a small fraction of the processing time available on their machines. So the SETI@home Project bundled up the processing software as a screen saver for personal computers and invited the public to take part in the research by downloading the software to their desktop machines.[9] What happens is that the software downloads a packet of data from the SETI central server and sets to work processing it. As a screen saver, the SETI@home software swings into action each time the computer is left inactive for a few minutes. The screen saver displays the progress of its analysis. When complete, the results are transmitted to the SETI server, which records the results and transmits the next packet of data for processing.

Now the processing on any given machine may be fairly slow, but by distributing the workload across thousands of computers, large volumes of data can be analysed very rapidly. By the beginning of 2001, SETI@home had attracted well over 2.6 million participants from 226 countries. Between them these computers had carried out over half a million years of processing time.

Other projects have emulated SETI@home's successful ploy. For instance, over a two-year period from September 1998 to September 2000, the PiHex project calculated the value of Pi (the ratio of a circle's circumference to its diameter) to over 14 million decimal places.[10] To do this PiHex distributed the calculation across 1734 computers in 56 countries and used up around 1.2 million hours of spare processing time.

The success of SETI@home and similar endeavours inspired many science projects to seek widespread participation. For example Casino-21, a climate simulation project, ran portions of the problem on participants' home computers.[11] Another project, Folding@home, aimed to distribute the calculations required to simulate the three dimensional folding of large protein molecules.[12] The Golem@Home Project aimed to use home computers to evolve novel robot designs.[13]

How to build a giant computer

How do you organise thousands of separate computers so that they work together to solve a problem for you? It makes no difference whether the different computers are different machines, spread across the Internet, or whether they are just individual processing chips packed inside a box. You still have to tell them all what to do. This was one of the first problems that computer scientists encountered as they began to build large parallel computers.

If every processor in the array does exactly the same kind of job, then the problem is relatively easy. You simply give them all the same instructions and pass out different parts of the datasets to each machine. This type of processing is called SIMD (Single Instruction, Multiple Data). We see this kind of processing in many real-life situations, such as the supermarket, where different customers (the 'data') are distributed among a row of checkouts (the 'processors'), which all operate the same way. Likewise, if you want a parallel computer (say) to add one to each of a thousand numbers, then all you need to do is pass out one number to each of a thousand machines. You tell them each to add one to the number they are given and send back the result.

Of course, no-one would bother spreading out such a simple job as adding one to a set of numbers. But suppose that each job was to run a simulation of the entire earth's weather patterns for a year, assuming slightly different levels of carbon dioxide and other greenhouse gases. Each such model might take a couple of hours to run, so if you were to do the job on a single machine it would take months. But spread over a thousand different machines, it could easily be completed overnight.

Most of the jobs that we saw earlier, such as SETI@home, were just like this weather example, but those are the easy cases. There are also many other problems for which we cannot carve up the processing in such a neat fashion.

One way in which this can happen is if the different processors need to share results with each other as they go along. Suppose that you want each processor to represent (say) a different car on a highway network, a different neuron in the brain, or the weather at different points on the earth's surface. The processors can no longer work merrily on their own. They have to exchange data constantly, just as cars on the road need to respond to actions of other cars, neurons in the brain signal need to one another and weather passes from one place to another.

One of the things that scientists need to know about systems like these is how the patterns of interactions affect the overall performance of the whole system (e.g. Figure 5.2). In each of the examples given above, the nature of the interactions is different from the others. But in many cases, the pattern of interactions are similar to one another. For instance, a forest fire is, obviously, very different from an influenza epidemic, which in turn is very different from a rumour. And yet, at a very deep level, all three of these phenomena involve a process that spreads from one element to another.

Similarities between different systems lead to the idea of *computational*

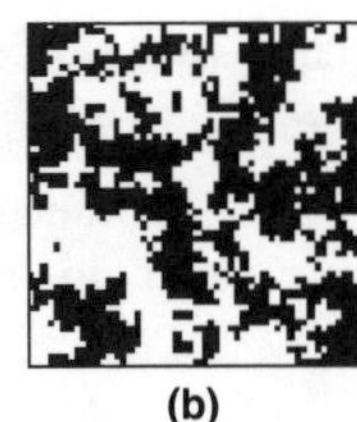

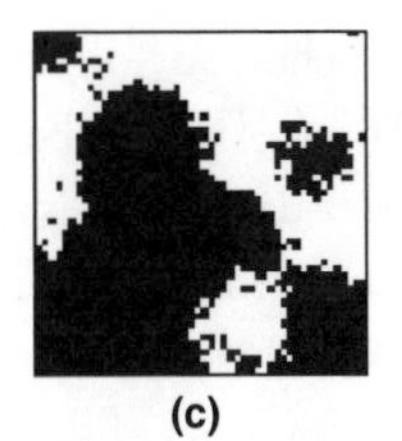

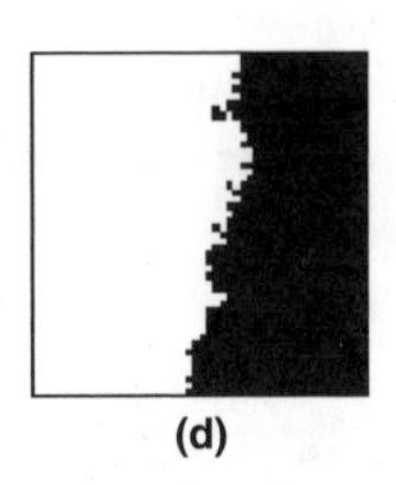

Figure 5.2 Using a cellular automaton model to demonstrate the effects of seed dispersal, fire and environment on the distributions of tree species Black and white represent competing kinds of trees. (a) Widespread seed dispersal results in random distributions; (b) short distance dispersal produces clumps; (c) fires make clumps coalesce into patches; and (d) geographic variation in environmental preferences leads to vegetation zones.

models, which capture the essential similarities between different processes. At first sight, a wildfire bears no relation to a 'flu epidemic. But when we look more deeply, we see that they both involve something spreading from object to object. Fire spreads from tree to tree, 'flu spreads from person to person. Computational models help scientists to understand how complex systems work. The point is that some of a system's features arise from its being a fire, a road network, or a brain, but others emerge from the pattern of interactions between its elements. By studying these interaction patterns, researchers can identify features that are common to many different systems.

It turns out that a relatively small set of computational models capture the essence of a very large number of systems. One of these models is the *cellular automaton* ('CA' for short). Scientists have built CA models of many processes in nature (e.g. Figure 5.2), such as the spread of fires and epidemics. To model weather, for instance, you divide the landscape into millions of cells, each covering a square area of the landscape. The result looks something like a checkerboard. Instead of having just black and white squares, however, the cells might be colour-coded according to the current state of the land surface, air temperature and pressure, and so on. The other important feature of the model is that neighbouring cells affect one another. Air moves, so the way in which conditions change in any one cell is affected by conditions in all the neighbouring cells.

By studying the behaviour of CAs, scientists have detected properties that are common to all of them. For instance, if the number of cell states is finite, and if the size of the grid is also finite, then the entire system must eventually settle down either to a constant, unchanging state, or else it will cycle through a set number of patterns forever. If cells can have an 'empty' state (for example, in a model of fire spread, this might correspond to a cell devoid of fuel), then the density of cells in this empty state must be below some critical threshold, or else the system becomes disconnected. That would lead to most of the non-empty cells becoming isolated from each other and behaving as individuals, with no interaction at all.

These and other properties of CAs are shared by any system that we can represent as a CA. They are universal properties that arise, not from the detailed biology or any other property specific to the system concerned, but from the way the model is wired together.

Several other computational models have widespread applications. A *Boolean network* (BN), for instance, is like an electrical switching network. It consists of a set of 'nodes' that are linked together in some pattern. Each node may be in either of two states, often represented as ON and OFF. Just as with CAs, the state of each node changes in response to changes in state of any nodes that are connected to it. Likewise, changes in its state affect the states of its neighbours.

One of the difficulties with large parallel computers is ensuring that different processors *can* talk to each other, a problem which confronted engineers as they tried to build the first processing arrays. For instance, as we saw earlier, the CM2 Connection Machine packed up to 65 536 processors into a single box. How do you ensure that each processor can 'talk' to every other processor? It is impractical to connect them all directly. For instance, to link together every pair of CM2 processors would require a total 2 147 450 880 connecting wires. In other words, the wiring problem rapidly dwarfs the processing problem. The wiring alone would fill an entire building.

Engineers faced exactly the same problem in trying to create computer networks. Even in a single office, with say 100 machines, 4950 direct wires are impractical. With millions of computers in the world, the problem becomes astronomical. The solution is to organise the connections as a hierarchy. Start with the office and its 100 computers. What you have to do is to connect all the computers to a single central computer, a network server. So if machine A needs to send a message to machine B, then it sends its message to the server, which then passes it on to B. Doing this requires no more than 100 connections, a much better option than 4950! In reality, various ingenious designs make it possible to wire the office with just a single loop of wire doing a circuit of the office.

To create a large scale network, what you do is to link the network servers for various local networks in their area to a domain server, which passes messages between local servers. This process continues on larger and larger scales. The ultimate expression of this approach is the Internet, which organises millions of separate machines into a hierarchy of domains.

Now suppose that machine A needs to send a message to machine Z on the other side of the world. To do this, A sends the message to its local server, which passes the message on up the hierarchy. It then descends down again into the domain where Z resides and the message gets delivered. This organisation is not as efficient as a direct connection, but it does get the message through and in no more than a certain number of steps. Most importantly, with a relatively small number of connections, this approach ensures that every computer can talk to any other computer.

Besides what they can teach us about the systems they represent, computational models have another practical use. They tell us what sorts of architectures are essential for large scale processing. In a CA model, for instance, each cell interacts with its neighbours. To model a CA, therefore, a parallel computer must be able to mimic that rectangular array, and the communication within it, as efficiently as possible.

The problems of connecting lots of processors together are hard enough, but they pale next to the problems that arise if you need each processor to do something different. To understand the problem, take a simple example. Suppose that you want to double a number and then add one, and you want to do this billions of times. Now suppose that you want one processor to do the doubling, and another to do the adding. So we pipe numbers through a pair of processors. The first one doubles the number; the second processor adds one to it. The following diagram illustrates this arrangement.

INPUT →	DOUBLE →	ADD 1 →	OUTPUT
item#4	item#3	item#2	item#1

This piping scheme speeds up the processing because it processes several different items (#1 to #4) at the same time (we assume for the moment that transferring numbers between processors takes next to no time at all). Factory assembly lines have used this principle for many years. So far so good. But what happens if we want to do a highly complex series of operations? If we spread the operations across a lot of processors the organisation of the calculations can easily become very complex. Instead of a simple sequence like the operation above, we might end up with a large network of operations, with many processes and pathways. Simply defining the pathways and processing can become an extremely difficult job in itself.

Because the organisation of parallel processing can become very complex it is essential to understand the properties associated with different sorts of 'wiring' and processing structures.

One advantage of computational models is that they allow us to study problems in the abstract. Earlier, we met several models of computation, such as cellular automata and neural networks. These models present ways in which the processors that are doing the actual calculations can mimic the real systems that they are representing. In a traditional computer—one with a single processing chip—the one processor does all the work. If the system it is modelling consists of many separate agents, then that single processor needs to compute the behaviour of each agent in turn. It first calculates the new state of every agent in the system, then it updates its internal representation of the system and repeats the procedure. As it computes the system step by step, it is building up a picture of the behaviour of the system in snapshots. So the model shows us the system's behaviour like a movie, frame by frame.

From the Internet to the Grid

When you switch on the lights in the evening, do you ever stop to wonder where the electric power comes from? By the end of the twentieth century, electric power could be generated by many sources—coal burning turbines, nuclear power plants, hydro-electric generators and wind farms—all of which feed power into an energy grid. In eastern North America, for example, this grid shunts power between states and across the border between the USA and Canada. The advantage of the grid is that the load can be diverted to areas where it is needed most, and constant resources such as the Niagara Falls need not be wasted when local demand is low.

If you want to move house, you do not go out and buy a truck; you hire a removalist who has the right equipment. If you want to put in a swimming pool, you don't buy an earth mover, you hire one. Likewise in computing, why spend $100 000 to buy a geographic information system if all you need is to draw a single map? One solution is to offer the resource over the Internet. This idea extends to many specialised kinds of processing that people might like to use, but would not buy the software for.

One of the grand visions of computing technology at the start of the new millennium is to create a grid of computing resources along similar lines to the power grid. Such a grid would offer the potential to create virtual organisations, in which groups with similar needs and interests can share resources. The emphasis here is on solving big problems, rather than building big computers, although the grid has been called a *metacomputer*. In effect, you are turning an entire network of computers into one giant machine. If you have a problem to be solved, you do not really need to know where the processing takes place. In a grid, a machine with a light load could take on more jobs from elsewhere. Also, the load of solving very large problems, such as climate models, could be distributed so that they can be completed quickly without monopolising the resources of any one computer.

The notion of a grid differs from the Internet in that the emphasis is on resource intensive applications, rather than data storage and delivery. It also differs from the idea of computing clusters, which normally consist of similar kinds of computers all administered under a single network. However, the potential applications of a grid can vary widely. First, there are applications such as simulation models that involve intense calculations.

Then there are problems in which jobs have a high throughput. The calculations required by SETI, for instance, are relatively small, but there is a constant stream of them. Finally, there are problems in which large volumes of data, possibly from many different sources, need to be processed.

A modern 'supercomputer' is more likely to be a grid of computers than a single machine. In grid computing, portals provide access to a range of computing services that can be accessed by other computers. The aim of grid computing is to integrate many different resources into a single, seamless whole. This goal poses many technical challenges, such as managing processing on different systems and passing data efficiently. Attempts at solving the problems involved in sharing resources began during the 1980s. The first large-scale grid experiment was a system called I-Way, developed in 1995. Perhaps the most influential system to date has been the Globus architecture, introduced in 1997 by Ian Foster at the Argonne National Laboratory and Carl Kessleman, from the University of Southern California.[14]

Where is all this leading? Eventually we will see many different sources and services integrating and sharing resources. Home appliances, for instance, are likely to share information and offer integrated controls. On a larger scale, we will see organisations with common interests sharing resources. These developments will make many new kinds of activities possible and their impact on our daily lives could be profound. The following two stories highlight both the potential advantages, and the dangers, implicit in these advances.

Happiness in Gridtopia

Frank Earnest is a modern company executive. His company provides him with all the mod cons. Today he is on his way to work in his company car. The car, which is on auto-pilot, downloads its driving directions from the local directory server and monitors its speed and position with the traffic controller, which coordinates traffic flow. Frank takes out his pocket computer and checks with his house server, which informs him that the refrigerator reports a deficit of milk and cheese. Frank approves the necessary purchases and requests that the house be at 22 degrees, with his favourite music playing by 7 p.m. Checking with his accounting service he sees that several bills have just been paid and asks that a deposit be paid to the resort where he plans to take his girlfriend at the weekend. Next he contacts the office server to check his schedule. Seeing that he has a meeting with a client at 10 a.m., he sets to work on a proposal and is just putting the finishing touches to it by the time he arrives at the office 15 minutes later.

To see what the flipside might be like, consider this nightmarish scenario.

Betrayal in Gridopolis

Frank is a freelance consultant who prepares and sells marketing reports to client companies. He is bidding for a new contract. Frank turns on his pocket PC to work on a confidential report. This contract is worth a lot of money and there are many competitors. Concerned about security, Frank wants to avoid using his usual word processing service, because he fears the transmissions may be tapped. So he downloads and installs an old-fashioned stand-alone program and goes to work.

Not being technically minded, Frank does not realise that this is a breach of the sales contract for his PC, nor does he realise that it automatically reports all such breaches to headquarters immediately. Software embedded in the downloaded operating system feeds Frank's data to the vendor, where it is scanned and onsold to a rival. By the time Frank submits his report, one of his rivals has already submitted his work under another name. When a legal service scans Frank's report, it detects the matching text and files a plagiarism suit against Frank. Responding to the resulting legal embargo, Frank's pocket PC shuts down to prevent its felonious owner from committing more illegal acts. Although the crime has been committed against him, Frank is the one who suffers. No-one planned it that way. It simply emerged from the system. He is the victim of the grid.

Computing in swarms

Many scientists, most notably Richard Feynman, have speculated on the possibility of building very small machines. Eric Drexler developed the idea into *nanotechnology*, which aims to build machines, even entire technologies, on a molecular scale.[15] Very small machines can do things that would be impossible by any other means. If we ignore the technical problems, then the possibilities are truly mind-blowing. For instance, *nanobots* (ultra-small robots) could revolutionise medicine. Placed in the bloodstream, swarms of nanobots could seek and destroy infections or cancers and repair damaged tissues. They could also revolutionise manufacturing: computer circuits of molecular size could be built directly; they could make ultra-pure substances by trapping impurities; they could repair fine-scale wear and tear on machines; create perfectly smooth surfaces; and assemble complex structures. The potential applications are almost endless.

For many tasks, nanobots would need some form of computer control. In a medical application, for instance, a nanobot would need enough 'intelligence' to identify and destroy a cancer cell, and to leave healthy cells alone. A swarm of nanobots would need to coordinate their activity. To do so, they would need to share information. The lessons learned about massive parallel computers, grids and the behaviour of large sets of processors will be crucial in nanotechnology.

chapter 6

NUGGETS OF KNOWLEDGE

Data expands to fill the space available for storage.
Parkinson's Law of Data

Knowledge and discovery

Up until the end of the nineteenth century, it was possible for any educated person to keep relatively well-informed of the latest theories and discoveries across the entire spectrum of science. Today it is difficult even for professionals to keep up to date in their own speciality. This blowout in the sheer volume of scientific knowledge has many consequences. Today it takes far more work to reach the frontier of knowledge in any area of science than it did even a generation ago. One result of this is that people tend to specialise more. They know more and more about less and less. Also, detailed scientific and technical knowledge is becoming increasingly opaque to outsiders—even scientists have a hard time understanding research outside their own area. It is not unusual for scientists in the same institute to be unable to understand each other's work.

Only a century ago, any well-educated person could understand the latest discoveries in biology or atomic physics. Today this is simply not possible. Increasingly, science is seen as an esoteric activity and this is almost certainly a large factor in the increasing public ignorance and mistrust of science, as well as its declining prestige.

Dealing with the sheer volume of knowledge is not new. It is one of the reasons why writing, books and libraries were first invented. The current problem has arisen because knowledge is now growing at an enormous rate and methods of handling it are not keeping pace.

New methods of handling knowledge have always stirred up resistance. One ancient Greek orator complained that the introduction of writing was making youth degenerate because they no longer practised the discipline required to memorise long speeches. When emissaries first showed a book to Shaka, warrior king of the Zulus, he could see no point of writing, let alone printing. Similarly, as an early advocate of Web technology, I found myself trying to convince one sceptical audience after another that placing information online was not a waste of time.

Similar problems keep cropping up in modern times. Students today learn to do arithmetic on a calculator. This frees them to concentrate on learning other, more advanced ideas, but the price they pay is less understanding of the processes. They become dependent on machinery, and are unable to detect mistakes. The same problem occurs with statistics packages. My first analysis of variance, for example, consisted of an entire morning with a calculator entering numbers and filling in the fields of a large table. Today the entire procedure is carried out by a spreadsheet at the touch of a button. Because it is now so easy, people often reel off advanced analyses that they do not understand fully, often drawing false conclusions.

The above examples highlight the need to understand the tools we use, the need to understand basic principles, rather than learning by rote.

The end of the expert?

A trend towards ever-narrower specialisation dominated science in the twentieth century. Driving this trend was the sheer volume of knowledge, which made expertise across a broad range of fields impossible. However, important discoveries today tend to occur at the interface between traditional areas of knowledge.

Narrow specialisation is no longer enough; different skills are needed. More and more scientists are temporary experts. They have a good general knowledge ranging across a wide range of fields, and deep knowledge about the particular problems and issues that they are currently working on.

Also, within the confines of school and undergraduate degrees, it is impossible to learn all the knowledge necessary to be a competent professional in a particular field. This problem has led to the proliferation of higher degrees by coursework, which blow out the training time still further. However, people cannot spend all their lives preparing for a career! Disciplines keep changing. If you spend ten years learning about information technology, then the content of your first few years' study will probably be out of date before you finish.

The only viable alternative is life-long learning. A basic education should give anyone enough to get started. We can then come back to further education as and when we need it. This includes refresher courses to keep us up to date with new developments. This issue is not confined to computing. For instance, the basic chemical skills that my father learned as a photographer are now giving way to the skills necessary to process digital images.

An important skill that people need today is to know how to learn. That is, to acquire new knowledge as and when they need it. We already see multi-skilling as a routine feature in the workplace. Canny employers have for some time included Web proficiency as an essential selection criterion.

Getting the message across

When Charles Darwin published *The Origin of Species* in 1859, he knew that something crucial—the mechanism of inheritance—was missing from his theory of natural selection. He never knew that the answer was found during his lifetime. In 1865, Gregor Mendel published an article setting out the basics of genetic inheritance, but this sensational discovery would lie unheralded, even unread, for 35 years. It was not until 1900 that three scientists, all working independently, serendipitously came across the article and saw its significance.[1]

Why was Mendel's seminal work on genetics lost to the world for 35 years? First, he published his results in an obscure journal—the *Proceedings of the Brunn Natural Sciences Society*. It is perhaps no accident that the scientists who rediscovered this work all lived in neighbouring countries—the journal would not have been available in the libraries of most other countries. A second, equally great obstacle was that most readers simply could not see the significance of the results. A casual reader of Mendel's article, which was titled 'Experiments on plant hybridisation', would see no connection with natural selection at all. Only when scientists began thinking about the role of inheritance were they able to make the crucial connection.

This is perhaps the most celebrated example of important results getting lost in the forest of scientific literature. Fortunately Mendel's paper was found and he was given proper credit as the discoverer of genetic inheritance. Many other scientists are not so lucky. Amid today's torrent of scientific literature, important results and ideas go unseen on a regular basis.

What convinced me of the enormous potential of the Web was an experiment I carried out early in 1993. Concerned that perhaps my best ever conference paper might never be seen, I placed a copy on the Web. To my amazement, more than 10 000 copies were downloaded within the first month. For authors, the Web's greatest advantage is instant worldwide distribution. However, there is still a potentially large gap between availability and our ability to access the information that we want. The story of Mendel's paper holds important lessons for the information age. On the one hand, the Web ensures that important information is available worldwide. On the other, there is a real need to organise that information so that crucial items are not lost like needles in a haystack.

The world's biggest book

In Japanese literature, anthologies of poetry are considered an art form in themselves. The idea is to select a number of short poems (*haiku* or *tanka*) and arrange them in such a way that there is an uninterrupted flow of ideas

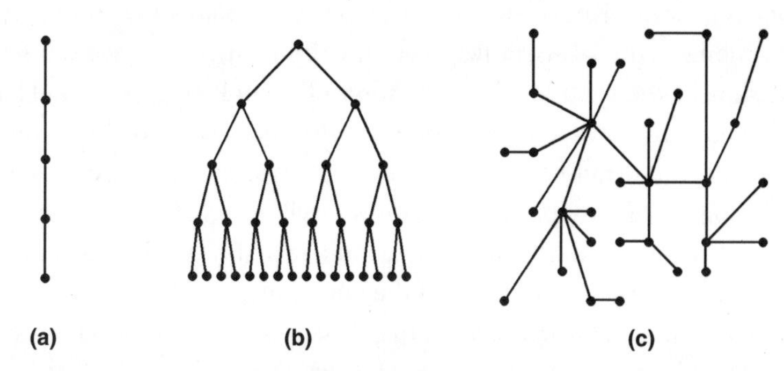

Figure 6.1 Three ways of organising ideas and information
(a) sequence; (b) hierarchy; (c) random. A traditional book adopts ordering (a); hypertext links on the Web can follow any of these patterns.

from one to the next, so that the whole is greater than the sum of its parts.

We can treat documents on the World Wide Web in a similar fashion. Suppose that we have a list of documents spread across many different Websites. Like the anthology of poems, we can draw them together to form an online 'book' about some common topic that we are interested in. All it takes is a table of contents with hypertext links from the list of 'chapters' to the documents concerned.

As with any revolution, the advent of the Web brings many traditional ideas into question. For example, what constitutes a book in the information age? The text may go through all the traditional hoops of publication such as quality assurance, but it never exists as a permanent whole. Part of the problem is that the Web frees publishers from the limitations of print. In a printed book, we are constrained to present ideas in a fixed, linear order and authors often face a dilemma about which order to use. Take poems, for instance. We might group our poems by poet, by chronological order, or by some flow of themes. The beauty of hypertext on the Web is that we need not be restricted to a single ordering of the information. We can include all the possible orders. All it takes is a separate set of links from one document to the next.

We need not even arrange information in linear fashion at all. Ever since the invention of writing, people have struggled to cope with the complexity of ideas. Knowledge is multi-faceted, ideas are interconnected, but traditional forms of narrative (speech, writing) are linear. We can see this tension in the traditional organisation of a book (Figure 6.1). The table of contents provides a hierarchical classification, with the linear text dipping into the detail of each topic in turn. The index provides users with random access to particular ideas as well as revealing cross-links between elements in different chapters and sections. In libraries, card indexes have traditionally performed the same function.

A thesaurus has a similar structure. In a thesaurus, one index groups concepts and words hierarchically; the other lists words alphabetically and gives links into the hierarchical index. This arrangement provides for an efficient compromise between the need for the listing to be concise and the desire to find references immediately. Every organisation confronts the same problem of information retrieval when storing records in their filing systems.

Rather than a linear flow, a Web 'book' can be a complex network of ideas. Each document can contain hypertext links from key terms and ideas to other documents that discuss them in detail. Moreover, these other documents can contain hypertext links to yet more documents. The documents themselves can be housed at many different sites and the chains of links from one document to another can go on and on indefinitely.

Such possibilities, combined with the explosive growth of information on the Web, sharpened the issues involved in organising and retrieving information. The first online indexes, handcrafted lists of links, were rapidly swamped by the gathering flood of items. Search engines were able to keep up to date by automatic harvesting of site contents, but the results often hid valuable grains in a mountain of irrelevant chaff.

Hyperlinks between elements allow users to browse through huge quantities of material, jumping from item to item while following a line of inquiry, or simply exploring. Also, because information is maintained at the source, it can be kept up to date and expanded without overloading the system.

To realise the potential described above, however, people need to organise online items of information. The most obvious approach is a top-down classification, much like a library catalogue, but classifications mask the rich associations of individual items. How, for instance, do you classify a document about Japanese poetry? Besides poems, it probably also contains useful historical and biographical information.

Why not just index everything? The current generation of search engines takes this shotgun approach and let the content of a document define its index. This approach has problems too. It means that a search will almost always throw up irrelevant items simply because they happen to contain the key words you are searching for. On the other hand, many valuable items can be overlooked simply because a specific keyword is not included.

Given the wealth of material online, bottom-up self-organisation is proving to be a more effective strategy. In other words, let the users create the order, just as ants unconsciously create order within an ant colony. The first step is to make information self-documenting and the simplest way to do this is to provide *metadata*, or data about data. You do this by adding in some basic details about the information when writing down your data or text.

In one way or another, we run into metadata every day. For instance, when a girl and boy meet at a party, the two things they need to know about each other are their names, and how they can contact each other. Whatever the material concerned, metadata must cover the basic context from which the information stems. Broadly speaking, metadata need to address these basic questions:

- HOW was the information obtained and compiled?
- WHY was the information compiled?
- WHEN was the information compiled?
- WHERE does the information refer to?
- WHO collected or compiled it?
- WHAT is the information?

Look at any book. Somewhere on its cover, you will nearly always find its name, the name of its author, and the name of its publisher. These pieces of information are metadata. Inside the front cover, you will find more metadata: the copyright date, the ISBN, the publisher's address, and more. When you are reading the book, or it is sitting on a bookshelf by your desk, these metadata are unimportant. But if you want to find it in a large library, which might hold a million books, then the metadata are crucial.

As it is for books, so it is for online data. Metadata are crucially important for storing, indexing and retrieving items of information from online sources. For instance, a document containing a proverb might look something like this.

WHAT	proverb
TEXT	Too many cooks spoil the broth.

This system sounds perfectly sensible, and it is. In 1998, librarians developed a standard called the *Dublin Core* for inserting metadata into Web documents.[2] However, metadata suffers from two disadvantages: most people ignore it, and some people abuse it. It takes time to add details explaining exactly what a particular item of information is. Many people do not understand the need for it, so they do not include it. A 1999 study found that only 34 per cent of Web pages included metadata; and only a fraction of those (about 0.3 per cent) actually conformed to the Dublin Core standard.[3] Commercial operations, on the other hand, which understand the value of promotion, also understand that metadata is a form of self-promotion. So they shamelessly add dozens of terms to ensure that their page of advertising will appear under as many different searches as possible.

Style versus substance

One of the many attractions of the Web is that people can design Web pages to look the way they want them to look. The underlying mechanics of this process is the *Hypertext Markup Language*, or HTML. This language allows authors to set text in **boldface**, *italics* and even in different colours. They can also control the layout of information on the page and create tables, forms and frames.

Controlling style is useful, but when distributing information, a deeper and greater concern should be to indicate the substance of the material. Suppose we create a document containing the well-known proverb

Too many cooks spoil the broth.

Using HTML, we could set it in italics by adding formatting tags like this:

> <I>Too many cooks spoil the broth.</I>

This is all very well, but suppose that someone is searching the Web for recipes. If they use the word *cook* as a search term, then the search will include our proverb, which is not about cooking at all. On the other hand, if they use the search term *proverb*, they will miss our document completely.

For this reason, it is important that a document provide clues about the sort of material it contains. We can do this by marking up the text according to *content*, not style. For example, we might wrap up our proverb like this:

> <proverb>Too many cooks spoil the broth.</proverb>

This new kind of formatting, called *content markup*, has several advantages. For a start, it is immediately clear what the text is. What is more, we can still get the document to look the same as before when it is displayed. All we need to do is to provide the browser with a *style sheet*, which will tell the browser to display proverbs in italics.

Content markup has many more advantages, one of which is that tags indicating structure also provide a convenient way of indexing documents. So a search for documents with the tag <proverb> would locate our document, but one that looked for the tag <cook> would not. In other words, content tags can greatly reduce the number of false returns in a Web search.

This system of content markup is called the *Extensible Markup Language*, or XML.[4] An associated standard, the eXtensible Style Language (XSL), is used to define the style associated with each tag. As we have just seen, XML is designed to help people to organise online material. XML also makes the content of Web documents much more flexible. For instance, suppose that a Website contained hundreds of reports about (say) business projects. To produce an overview of the reports, the system might simply lift the sections *title*, *author* and *summary* out of each individual report and collate them into a single overview document.

In reality, using XML is not quite as straightforward as the above description suggests. For instance, the content tag <summary> could appear in many contexts, so there is still plenty of scope for errors. One answer is to use different languages—different sets of tags—for different kinds of knowledge and activity. Chemists need to identify documents that deal with chemistry, geographers need to identify files containing maps, and so on. As a result, many specialised markup languages were introduced, each with its own set of special purpose tags. Examples include MathML, MusicML, ChemML, for dealing with the specialised notations used in those three fields.[5]

These specialised languages allow us to link together related material from many different sites. Geographers, for instance, are developing systems for providing geographic information over the Internet.[6] Many online services already provide maps of geographic information, such as the locations of tourist sites in a region. The ultimate expression of this would be a worldwide mapping service that enables people to produce maps on demand for any part of the world, and showing *any* features that they want or need. If you want a map of Highland lochs to plan your up-coming fishing trip, then such a system should, in theory, be able to oblige.

However, building such a system is not so simple in practice. Each agency or company that provides geographic information usually deals with only a limited geographic region, and only certain features of that region. The Scottish Parks authority, for instance, would deal only with parks in Scotland, not Ireland, and might not handle information about many other features, such as hotels or mineral resources. You may have to gather together details from several different sites on the Internet in order to produce the map you want for your fishing trip. Making sure that this is possible requires a great deal of coordination. It is essential, for instance, that everyone provides their information in a consistent fashion.

At the time of writing, many different groups and organisations are working feverishly to develop interchange and markup standards for all kinds of data. This is largely a response to the online environment and the proliferation of XML, and XML-based languages such as the Geographic Markup Language (GML) and the Data Space Transfer Protocol (DSTP). It is crucial to achieve consistency between different authorities. For example, the Species 2000 Project is developing standard XML namespaces for biodiversity data.[7] However, these data include geographic elements, so the compilers had to use conventions that made it possible to merge biodiversity data with other kinds of spatial data.

A virtual warehouse

The standard way of storing electronic data is in a database. In Chapter 3, we saw some of the methods used to design databases. In today's information-rich world, however, there are many databases of many kinds. More and more, people are seeking to exploit the serendipity that occurs when you combine different databases. Combine weather data with data about hospital admissions and you may find clues about the effect of weather on road accidents. For this reason, governments, researchers and companies have begun to compile vast storage facilities for data. Too big to be called simple databases, these facilities have become known as *data warehouses*.

A data warehouse consists of many different databases, which gives rise to many practical problems. For instance, different datasets may not meet the same standards. Legacy data, for instance, is old information that is still useful but it is likely to be stored in idiosyncratic formats and may require special software to extract the data. For instance, during a project to compile biodiversity data, we wanted to use latitude and longitude to indicate the geographic location of each record. However, many old records referred to coordinates on old survey charts, so the values had to be converted. Even more challenging were records in which location was given by a description such as '5 miles north of Gundagai'. Converting these records required software that could read the description, refer to a gazetteer of place names and then make the necessary calculations.

To cope with such problems, data warehouses often embed the original databases within *information objects*, which hide the messy details of how the information is extracted and converted. We can then use these objects as building blocks to create sophisticated information resources. A map, for instance, can be treated as an object that is built out of many sub-objects. There is no need for all the components to sit on the same computer. A distributed data warehouse (DDW) can be built out of information objects from many sites on the Web. Authorities such as the World Wide Web Consortium (W3C) have developed standards to encourage the development of DDWs.[8]

Until recently, most research was carried out as a series of separate studies. However, the Internet allows the outputs of past research to enrich many more subsequent studies. A good example is genomic research. Large, online databases (see Chapter 9) make it possible to compare new sequences with whole families of existing data. They also enable new kinds of studies, such as looking for unsuspected patterns and relationships. One of the main motivations for creating the Internet in the first place was the potential to share and combine information from many different sources.

Some of the first online warehouses were not concerned with data at all, but with software. Just as the US military played a key role in the early development of the Internet, it likewise had a hand in early efforts at data warehousing. The US Air Force set up a gigantic online library of software, called Simtel, originally located at the White Sands Missile Base in New Mexico. It stored thousands of freeware and shareware computer programs that anyone could download and use. The site was so popular that dozens of other sites all around the globe soon created 'mirror' sites to help meet the demand. (A mirror is a complete copy, like a mirror image, of a service at a second site.) In this network of mirrors, each site would regularly download material from the original host to keep its copy up to date. The original Simtel site closed in 1993, but the service it provided continues through the system of mirror sites.

The Internet (especially the World Wide Web) makes it possible to combine information from many different sources in a seamless fashion. This leads to data sharing and cooperation on scales that were formerly unthinkable. At first, the explosive growth of the Internet led to confusion—many organisations duplicated facilities in inconsistent ways—but by the turn of the millennium, there were concerted efforts underway to develop protocols and standards governing such issues as data recording, quality assurance, custodianship, copyright, legal liability and indexing.

From interpretation to data mining

As people began to gather more and more data, they began to wonder what they could do with such large amounts. As we saw in earlier chapters, the serendipity effect makes itself felt in large collections of data. If the collection is large enough, then you don't need to be particular about what data you have. There are bound to be unexpected patterns and relationships hidden within it.

In both science and business, researchers have begun to exploit the serendipity effect to detect useful relationships hidden within large datasets. This important new research tool, known as *data mining*, has proved to be an extremely fruitful way of uncovering unsuspected, and often profitable, 'nuggets' of information. It draws on many existing areas of computing and mathematics, including databases, statistics and machine learning.

The information explosion occurred so rapidly that methods for interpreting the data struggled to keep pace. What I might call the 'traditional' methods of interpreting data are based around the assumption that there are not many data to interpret. What is more, they mostly assume that you have collected the data with the aim of answering a particular question. Data miners have therefore been busy developing new ways of exploring data. Data mining is a highly inter-disciplinary area of research: it draws on technologies from database technology (data warehousing), machine learning, pattern recognition, statistics, visualisation and high-performance computing.

Data mining is popular because computerised trading generates commercial data in vast quantities. For example, cash registers can now record and analyse every purchase. Even a single day's trading at a suburban supermarket can involve tens of thousands of separate purchases. Any single purchase may not tell us very much, but there may be very revealing patterns lurking in the log of a day's trading. The combinations of products that people buy may give the store manager hints about how best to lay out the shelves. Day-to-day changes in the sales of particular products may indicate whether marketing strategies are effective.

In some respects, the term data 'mining' is a misleading analogy. Conventional mining begins only when an ore body has been located and proved to be commercially viable. The first step is minerals exploration. So data mining really begins with data prospecting. You start mining the data in earnest only after you have found a useful vein that you can exploit.

The first thing you need is to find useful patterns and relationships within your data. The point about data exploration is that you never know what you are going to find. You have to explore the data, looking for any 'nugget' that you can use. Nuggets can take many forms. Often they are relationships, such as 'potato chips and cheese dip are often sold together', 'the volume of sales is always higher on Friday', or 'there is a significantly higher incidence of cancers in low-income areas'.

Perhaps the most famous story about data mining concerns an odd coincidence that was found by a study of sales records for a grocery chain in the American midwest. The study revealed that on Thursdays many men would visit their local shopping centre and buy just two items: disposable nappies and beer. It turned out that many young families with infants lived in the area. Although they would do their shopping at weekends, nursing mothers with young babies would often run low on nappies towards the end of the week. So they would ask their husbands to buy some on their way home from work. The husbands would take the opportunity to pick up a supply of beer for the coming weekend.

Data mining is such an effective commercial tool that companies are constantly looking for ways to gather more and more information about their customers. Reward schemes which award points for purchases are really ways of encouraging customers to provide more data about their buying habits.

Of course a lot of people will be worried about the potential for invasion of privacy raised by the collection of so much data about almost every activity. In many instances, steps are taken to hide the identity of the individual. However, even if obvious indicators such as a person's name and address are removed from records, there is still a risk that there may be enough other clues to identify that person.

Serendipity for profit

Data mining exploits the serendipity effect in a big way. The starting point is the assumption that interesting and useful relationships lie buried within the data. As we saw in Chapters 1 and 2, when you throw different kinds of data together, interesting and unexpected discoveries are inevitable. The challenge is to find them.

In traditional research, data analysis is the final icing on the cake. Normally it consists of a single analysis to test a single hypothesis. In contrast, data mining routinely tests hundreds, if not thousands, of possible relationships. Like a musical virtuoso, the computer runs through variation upon variation on a theme, until it strikes one that looks promising.

This is a relatively new way of doing research which came to the fore during the 1970s as the power of computers increased. However, many older scientists were (and still are) uncomfortable with the idea of deliberately setting out to make scientific discoveries without knowing what you are looking for. This is surprising, because science has always exploited serendipity, as we saw in Chapter 1.

An important issue for data mining is a problem known as *combinatorial complexity*. As we have seen, the number of ways in which you can combine different pieces of data literally explodes as the size of the dataset increases. So a lot of the research in data mining is directed at new and innovative ways of coping with the combinatorial explosion.

To get a better feeling for some of the issues look at the simple example of a medical database in Table 6.1.

The table shows several records of the kind that a health agency might obtain in the course of a medical survey. The data are laid out in typical fashion. Each row provides what is known as a *record*, the data for one particular person who has been surveyed. Each column represents values for a particular variable or attribute across all people surveyed. Some of the columns have numbers as values (*Age–Height–Weight–Children*),

Table 6.1 A table of results from a hypothetical medical survey

Name	*Sex*	*Age*	*Height*	*Weight*	*Job*	*Married*	*Child*	*Smoke*	*Heart*
Anne	F	32	163	59	Teacher	0	0	Yes	No
Bill	M	25	181	73	Plumber	0	0	No	No
Fred	M	56	177	77	Sales	1	3	Yes	No
George	M	43	175	91	Manager	1	2	No	Yes
Gladys	F	47	167	52	Doctor	1	3	No	No

some have text (*Name–Sex–Job*), and the rest take the binary values 'Yes' or 'No' (*Married–Smoke–Heart condition*).

Let us suppose that we have in mind a definite question that we want answered. For instance, we might want to ask the question 'Is *Weight* related to *Height*?' To answer that question, we could draw a plot of the column for *Weight* versus the data in the *Height* column.

Another typical question we might ask is whether men weigh more than women. To answer this we need to divide the data into two parts on the basis of the sex of the individuals. We can treat other binary variables in the same way. So, for example, do people with heart conditions weigh more than other people? There are specialised techniques for detecting and testing relationships that involve numeric variables, and likewise for non-numeric variables. For instance, we could test the validity of rules such as

IF *Sex* = Female and *Smoke* = No THEN *Heart* = No.

Identifying this rule amounts to testing the hypothesis that non-smoking females do not have heart conditions. Now we have to be careful here. This hypothesis certainly does not hold true in the general population, but it is true of our small sample. This is why large datasets are so useful. If you identify something that is true, or even statistically significant, then the chances are that it gives you a useful insight about the general population.

Another operation that we can carry out is to classify parts of the data. For instance, we could define cut-off values to classify people as *Young*, *Middle-aged* or *Elderly*. Likewise, a person might be *overweight* if their weight is greater than some threshold value determined by their age and sex.

What is most important is that even within this tiny data set, there are already many different questions that we can ask, each of which has the potential to reveal important insights. Not only can we look for relationships between different combinations of variables, but we can also restrict them to different subsets of the data. So we could look for a relationship between *Weight* and *Height* for men only, or for married people, and so on. Even with a small set of data like this one, we could ask literally dozens of questions.

Strictly speaking, our sample table is just a bit too small. I have deliberately kept it short for reasons of space. A real survey might easily contain thousands of records, and dozens of variables. Many of the methods that data miners use simply do not work with small samples like this one, as we saw in the case of the hypothesis about non-smoking females. Nevertheless, the table can serve to show that even with such a small set of data, there are still many ways in which to look at the data.

In real databases, the number of possible ways to carve up the records and variables quickly grows astronomical. If we explore the various permutations and combinations, sooner or later we are bound to uncover a significant relationship. This does not have to be a hypothesis that we deliberately set out to test, but rather one that simply pops up from the data.

These relationships are the 'nuggets' data miners are searching for. The analogy with gold prospecting is apt, for these nuggets of knowledge can sometimes be worth a fortune in gold, literally. Data miners want to make sure that they find those nuggets. Like a needle in a haystack, finding a particular relationship may not be easy and, for this reason, data mining studies are often carried out on supercomputers. Coming up with more efficient algorithms to search for patterns is a high priority for research in this increasingly active field.

What sorts of nuggets do people look for? What sort of information could possibly be worth a fortune? One extremely active area is in marketing. Companies spend billions of dollars each year in advertising and promoting their products. Anything that helps them to do the job more efficiently can not only save them a mint in advertising, but also increase sales.

Data mining includes many kinds of data processing. As we saw above, one activity is to find rules of association, but perhaps the most common activity is cleaning up data. When you combine records from different sources, there are often records with missing, irrelevant, incorrect or incompatible data items.

Another activity is to characterise particular items of information. For instance, if you can build a profile of the kinds of people who buy your product, and if you can identify people who fit that profile, then you can target your advertising at your most likely customers. A related activity is to decide which class different records belong to. For instance, is an applicant for a loan a high risk or low risk? When there is no obvious way to classify information, you can find clues by clumping similar records together. What groups of customers are being admitted to hospital, for instance?

All of these questions have commercial applications. In the highly competitive world of big business, any tool that improves efficiency can increase profits dramatically. In a multi-billion dollar a year industry, for instance, even a 1 per cent improvement represents tens of millions of dollars. In the next two chapters we look at new technologies that have applications to data mining, followed in Chapters 9 and 10 by applications in biotechnology and environmental management. One of the keys to data mining is automation. We now turn to a new kind of automation that is playing an increasingly important role in e-commerce—intelligent agents.

chapter 7

TALK TO MY AGENT

For knowledge itself is power.

Francis Bacon

The Knowledge

Since the time of Queen Victoria, London cabbies have had to qualify for their permit by passing a stringent test of their ability to navigate London's streets and locations. Called 'The Knowledge', the test requires cab drivers to know exactly how to find their way from any given location to any other. London is a big city. Add up all the streets, hotels, churches, stations, stores, and so on, and you get literally thousands of individual locations. But that is only the start of the problem. You need to know how to travel from any of these thousands of places to any other, so there are literally millions of possible trips. And cab drivers must know them all!

How do people ever manage to learn all these millions of possible routes? The answer is that they do not. What they do is learn a set of standard routes. The so-called 'Blue Book' put out by the Public Carriage Office contains 400 basic 'runs'. Every possible trip then becomes a variation on one of these standard runs. If a hotel is next to a church and you know a route that passes by the church, then you know how to get to the hotel. If your destination is in a remote side street, then all you need to do is to identify which of your standard routes passes nearby.

People use similar tricks to solve problems. Our lives are full of recipes for doing things: how to recognise faces; how to make a pot of tea; how to entertain guests; how to drive a car. The list is endless. And just like the knowledge that the cabbies have to learn, there is no way that we can learn everything we ever need to know by rote. So, what we do is to learn a lot of recipes. We then adapt those recipes as we need to. For instance, learning to draw provides kids with basic skills, such as holding a pen and coordinating hand and eye movements. Those skills can then be applied to lots of other tasks later in life, such as writing, or using a mouse. The most important knowledge is the most basic—the things that we learn as babies or as toddlers, such as how to eat, how to walk, how to talk, and so on.

Doing is knowing

Concepts without factual content are empty; sense data without concepts are blind . . . The understanding cannot see. The sense cannot think. By their union only can knowledge be produced.

Immanuel Kant

In dealing with the world around us, there is a hierarchy of understanding: data, information, knowledge, wisdom. The most basic elements are data—raw facts about the world. 'The sky is blue.' 'The sun rose this morning.' 'A male gorilla weighs 300 kilograms.' And so on. Information is data that has been distilled, so that essential patterns and relationships become clear. 'Shops are closed on public holidays.' 'Gravitational attraction decreases as the inverse of the square of distance.' 'Overweight people are more susceptible to heart attacks.'

Knowledge goes one step further. It consists of details of how to use information. 'Keep left when driving on English roads.' 'Take an umbrella when you go out on a rainy day.' 'Buy shares when the price is low, sell when the price is high.' Finally, we come to wisdom. Wisdom is the ability to gather knowledge and adapt it to new situations. In other words, wisdom is knowing when, where and how to apply knowledge. It is knowledge at a deeper level. 'Always be firm but fair.' 'Be flexible.' 'Keep an open mind.' 'Avoid hasty decisions.'

A sound knowledge of the world around us is indispensable. It is not enough to learn lots of facts or to acquire information. You need to be able to respond to those facts and to use that information. One of the great challenges for artificial intelligence research has been to devise intelligent computer programs that can acquire and use knowledge.

The essence of knowledge is the ability to use information to help you do things. This is usually expressed as a 'rule'. In general we can express a rule in the form 'If A is true then B is true', or 'If A happens, then you must do B'. We use rules all the time, although most of us do not realise it. The following simple examples show what some rules might look like in different contexts. In games like chess, we use both the rules of the game, as well as useful rules about tactics and strategy.

If the king is threatened, then
if the king can move,
then move it
else if you can take the threatening piece,
then take it,
else resign the game.

We also use rules to classify things. Here are some examples drawn from several different contexts.

IF height > 6 feet
THEN person_type = tall

IF manufacturer = Toyota
THEN country_of_origin = Japan

IF person_age ≥18
THEN status = adult ELSE status = juvenile

IF land_cover = trees
THEN area_type = forest

Most computer programs incorporate rules, either explicitly or implicitly. Electronic mail programs, for instance, allow the user to filter incoming mail. They may automatically place copies of messages in different folders based on words in the subject line, or on the sender's address. The typical format is a table indicating which part of the message to look at, what contents to look for and what to do whenever the conditions are met. Here is an example:

Field	*Contents*	*Action*
sender	junk@mail.com	Delete
subject	'Canada project'	Copy to folder

Word processors use rules too. Automatic spell checkers have lists of common typos and their replacement strings. For instance, the automatic correction feature on the word processor I am using will correct such typos as 'adn' for 'and', 'teh' for 'the', as well as substituting symbols, such as '©' for '(c)'.

One kind of program, an *expert system*, is designed specifically to use and manipulate knowledge. Expert systems often contain hundreds, or even thousands of such rules. We can consider the rules as links in a network of concepts. So A is linked to B if the *knowledge base* contains a rule such as 'IF A THEN B', which we can abbreviate as $A \rightarrow B$.

The usual search mechanism is a procedure known as *backward chaining*. This approach starts from the statement that it needs to satisfy and works backwards. For any potential solution X, it tries to form a chain of rules

$$P \rightarrow Q, Q \rightarrow R \ldots, Z \rightarrow Y, Y \rightarrow X$$

in which the conclusion (right-hand side) of one rule is the precondition (left-hand side) of the next rule. The program continues this process of forming chains until it finds a chain that starts from a condition that it can check directly and ends with the statement it is trying to prove.

One of the outstanding features of expert systems is their ability to use chaining to search relentlessly through the network of rules for the answers to complex queries. The knowledge base for a family tree would consist of numerous facts, such as:

'Adelaide is the mother of Helen.'
'Mary is the mother of Adelaide.'
'Mary is the mother of Frank.'
'Frank is the father of George.'

It would also contain knowledge about how to define certain kinds of relationships, such as:

'X is sibling of Y if Z is mother of X and Z is mother of Y.'
'X is grandmother of Y if Z is mother of Y and X is mother of Z.'

Given a goal such as 'find all siblings of Frank' or 'find all cousins of Helen', an *inference engine* would trace its way through the rules, systematically trying out every combination until it had exhausted all possibilities.

The rules of the game

Recognition of realities is the beginning of wisdom.
Inscription on the tomb of Juho Kusti Paasikivi (1870–1965), former president of Finland

How can you test whether or not a computer program is intelligent? This question has challenged scientists ever since the invention of the computer. Frustrated by the difficulty of pinning down just what intelligence *is*, people often turn to a simple test suggested by Alan Turing, one of the pioneers of computing. *Turing's Test*, as it is known, can be stated as follows:

> If a computer can perform in such a way that an expert cannot distinguish its performance from that of a human who has a certain cognitive ability then the computer also has that ability.

Imagine the following scenario. Suppose that you go into a room where you find a computer that is connected to a similar machine in another room. You type questions and comments into the computer at your end, and responses come back from the other room. But who is sending back the replies? Is it a person in the other room? Or is it a computer? Now suppose that it is really a computer program, but you cannot tell whether or not it is a human who is answering. Under such circumstances Turing's Test would assert that the program is exhibiting intelligence.

In Ridley Scott's movie *Blade Runner*, Harrison Ford plays the role of a policeman whose job it is to track down replicants—robots that imitate humans almost perfectly. The technique he uses is an application of the Turing Test. He simply asks the subject a series of questions, probing ever deeper until the robot reaches the limits of its artificial memories and starts behaving oddly.

In recent years, a number of competitions have sprung up in which different programs are pitted against one another in a Turing Test. The program that fools the greatest number of people wins.

One drawback of the Turing Test was highlighted in the 1980s by the program ELIZA, which was designed to embody simple ideas from psychiatry.[1] ELIZA was designed to draw out discussion by extracting terms from the user's comments and throwing them back as questions. It also throws in random ideas as well. Here is a typical conversation:

ELIZA: Hi, what is your name?
User: Fred.
ELIZA: What is your problem, Fred?
User: The weather.
ELIZA: Go on.
User: It's very hot lately.
ELIZA: Why is it important that it is very hot lately?
User: Well it's been over 30 degrees every day for a week.
ELIZA: What if it were otherwise?
User: Then I'd be happy.
ELIZA: Is it appropriate to be happy?
User: Yes.
ELIZA: Tell me why.
User: That's what my parents taught me.
ELIZA: Tell me more about what your parents taught you.

The ELIZA program had no real intelligence at all, just a few simple conversational tricks. And yet for all that, many people felt that it was truly intelligent. Some users even became addicted and carried on long conversations with the program!

The point is that ELIZA *appears* to act intelligently without understanding anything that is said to it. The interaction of the user's inputs with a few simple rules (e.g., 'throw their words back at them') can produce complex conversations that many would call 'intelligent'.

From expert to agent

In almost any area of activity, people refer to experts for advice. It may be the sports coach, the professional handyman, the TV repairman or the wine maker. Not surprisingly, people have often tried to reproduce some of the knowledge of such people in computer programs. Called *expert systems*, these programs are usually developed to provide instant advice in areas where many people need help. Simple examples include such everyday matters as selecting a colour scheme for your bedroom, diagnosing a fault in your computer, or choosing a holiday.

In building an expert system, the human expert tries to incorporate all the knowledge that is necessary to solve a particular kind of problem before the system is used. This may involve a long development procedure and it works best when the knowledge domain is relatively narrow. For instance, an expert system that diagnoses faults in your TV set is more likely to be comprehensive and reliable than one that tries to diagnose medical ailments.

There are many kinds of problems for which complete knowledge is unobtainable. In such cases, acquiring new rules and data needs to be an on-going activity of the system. For example, on the Web, new sites are always appearing and new data are constantly being added to existing sites. Systems must be able to add details of such items to their information bases regularly.

An expert system differs from a database in that it includes the knowledge needed to solve a problem. Basically, you query a database, but an expert system queries you. The sample rules for taxonomic classification presented in Chapter 5 form a good example of a simple kind of expert system known as a *decision tree*. In practice, it would present the questions

'Does it have sharp teeth?', and so on one at a time, gradually guiding the user to the name of the animal. (A real taxonomic expert system would be more sophisticated than our example!)

Fully developed expert systems are relatively rare. However, as we saw earlier, the idea has inspired a growing use of knowledge in programs of all kinds.

One of the most talked about ideas in computing today is that of an agent. An agent is a program that represents your interests in some way or other. The way they represent you is usually very narrowly defined as a set of goals. People use the term 'software agent' to mean many different things, so the precise definition varies enormously. However, features commonly associated with agents include the ability:

- to use knowledge and intelligence;
- to learn new skills;
- to interact with other agents;
- to take independent action, without waiting for directions from the user.

One sense of the term *agent* covers programs that act on behalf of a user. For example, most Web search engines use software agents that automatically trawl through Websites, recording and indexing the contents. A related definition concerns programs that automatically carry out some task or function. The term *adaptive agent* refers to software that changes its behaviour in response to its 'experience', that is, the problems that it works on. In general, the software agents used by search engines are not adaptive. Although they accumulate virtual mountains of data, the way they function is essentially unchanged.

The potential of agents goes a lot further still. Perhaps most exciting of all is the possibility of agents talking to each other. Rather than acting alone, your agent might talk to agents representing (say) your friends, colleagues and organisations that you wish to keep in touch with. This prospect raises the question of what sorts of social organisations might emerge from such practices.

Such questions are by no means trivial. One of the most important insights provided by the new research field of artificial life is that interactions between agents can have profound effects. For instance, in one early study, Paulien Hogeweg and Bruce Hesper showed that the observed social organisation of bumblebees arises as a natural consequence of the interaction between simple properties of bumblebee behaviour and their environment.[2] One rule they invoke is the TODO principle. Bumblebees have no intended plan of action, they simply do whatever there is to do at any given time and place. Similar interactions lead to order in many other animal communities, such as ant colonies, as well as the formation of flocks of birds, schools of fish and herds of antelopes.

Online agents

The Internet gave a great boost to agent technology. Suddenly there was an entire world of information for agents to inhabit and explore. One of the first applications was in indexing online resources. The major Internet search engines have employed agents for years. They go by many names, such as crawlers, spiders and walkers. The role of these 'virtual robots' is to locate and record details of items provided on Internet sites around the world.

A Web crawler is a software agent that trawls the Web recording and indexing everything it finds. Mostly, it is looking for metadata—the kinds of data about data that we met in Chapter 6. Later developments have included collaboration whereby a single query can spawn searches on many different sites, after which the results are pooled and transmitted back to the user.

Agents provide one possible solution to the problem of coping with the increasing flow of information that we saw in the previous chapter. For instance, instead of you having to read the thousands of messages that can pour in from mailing lists on a daily basis, your agent could peruse the messages for you, drawing to your attention those that contain key phrases, come from particular sources, or deal with particular topics. This does not eliminate all the irrelevant messages, but it can reduce the sheer volume of messages by orders of magnitude. Extending this idea, a 'news agent' could peruse online information for you. It could open up newspaper sites online and search current stories for items that might interest you.

As with most technologies, online agents pose potential threats as well as opportunities. As the number of search engines increase, Web managers are becoming alarmed at the intensity of hits generated by robot agents as they sift through the contents of their site. If personalised Web crawlers became a fact we could easily reach a state where the entire Internet is engulfed by a robot feeding frenzy.

Unfortunately, one kind of agent has attracted more attention than any other: computer viruses. Computer viruses have all the characteristics of agents. They act independently. They reproduce themselves. They move from one computer to another. But, unlike other agents, they cause mischief. Though they do not learn in the normal sense, they certainly do explore, by finding ways to spread. In recent years, email attachments have become the greatest carriers.

As if the problem of computer viruses is not already bad enough, there also is the bleak possibility of new viruses appearing spontaneously. This fear is sparked by the increasing number of mobile agents, pieces of software that can move around the Web from machine to machine. They often take the form of Java applets, called 'AGLETS'. One concern is that copying errors might corrupt normally harmless software and turn it into a virus. The danger is greatest with agents that already replicate themselves, or that are regularly copied. However, such a danger is infinitesimal compared to the host of contaminated email attachments that now fly around the Web every day.

Sharing and learning

If agent software is to be intelligent and to act independently, then somehow it must learn the necessary skills. Obviously, these can be programmed by their users, but how can they acquire knowledge independently?

One way for this to happen is for one agent to pass on knowledge direct to another agent. How can two agents share their knowledge?

One option is for agents to pick up pieces of knowledge that are deliberately left lying around for them to use. For example, in August 2000, the Data Mining Group (DMG), a Chicago-based consortium of academics and industries, released the *Predictive Modelling Markup Language*

(PMML). The press release described it as 'an XML-based language which provides a quick and easy way for companies to define predictive models and share models between compliant vendors' applications'.

PMML allows people to describe different models in a standard way. This means that a general purpose software agent can simply go and read up on what to do each time you give it a new task to perform.

Another way for agents to learn is by assimilating knowledge from their own experience. That is, they add new rules to the repertoire of things they can do. This idea is perhaps best illustrated via an example. Suppose that to carry out a particular kind of Web search, the user has to feed the agent with a set of explicit instructions. Not only could the agent carry out the query, but it could add the instructions to its knowledge base. This would allow future users to activate the same query without needing to instruct the agent how to go about it. Of course, to ensure that future users know that such a query is available, it needs to have a name by which it can be invoked.

The Knowledge Query and Markup Language (KQML) provides a protocol for exchanging information and knowledge.[3] A query written in KQML amounts to knowledge of how to find the answer to a particular question. One agent can use it to request information from another agent, or two agents can exchange what they know, either directly or by creating knowledge depots.

Suppose that we ask an agent to provide a report about kauri trees in New Zealand. To carry out the query, we give the agent an appropriate set of rules, and we supply a name, NZKAURI, for the query. This query then becomes a new routine that future users can recall at any time simply by quoting its name. However, it is limited by its highly specific restriction to forest trees and to New Zealand. We can generalise the query by replacing the specific terms with generic variables. But how do we generalise terms like 'forest trees' and 'New Zealand'? Suppose that we express the query using XML. To generalise the query, all we have to do is replace specific references such as 'kauri' and 'New Zealand' with the variable fields <tree> and <country>. This turns the query into a function TREEPLACE (*tree, country*) (see Table 7.1).

This function now has the potential to address questions on a much broader basis. In principle, it could answer questions about (say) kauri

Table 7.1 Turning a query into a function

Original query	*Resulting function*
<query name = 'NZKAURI'>	<function name = 'TREEPLACE'>
<tree>kauri</tree>	GET <tree>
<country>New Zealand</country>	GET <country>
. . . rest of procedure . . .	. . . rest of procedure . . .
</query>	</query>

trees in other countries, other trees in New Zealand, in fact any kind of tree in any country.

The success of such generalisation depends on how queries are implemented. For instance, if the query uses a number of isolated resources that refer only to New Zealand kauri trees, then it will fail completely for any other tree, or for any other country. The generalisation is most likely to be successful if the entire system confines itself to a narrow domain.

From ants to agents

So far we have looked at ways of applying biological ideas to develop better methods of computation. In contrast, new fields of research, such as Artificial Life (Alife for short), attempt to understand biological processes by replicating them in computer models.

In Chapter 5, we looked at the ways in which people harness the independent efforts of many different computers to solve large problems. In all our examples, however, the individual computers never interacted with each other. They simply took their orders from a central server and reported their results back to that same machine. But it is also possible for different computers to pool their activities in ways that require direct interaction.

Coordinating the activities of perhaps millions of separate computers is far from trivial. The more computers that are involved, the greater the potential for monumental foul-ups! When the individual machines act independently, coordination is relatively straightforward. You just keep track of what each one is doing and record the results. But once they start interacting with each other, all sorts of problems can arise, such as unnecessary duplication of effort, or large numbers of machines standing idle while one machine finishes its current task.

Computer scientists are looking increasingly to the natural world to learn how to resolve potential problems. Living things have evolved ways to solve all manner of complex problems. As a result, some areas of advanced computing have become almost indistinguishable from biology.

In the book *Patterns in the Sand*, I described the way in which ant colonies emerge out of millions of simple interactions between ants and each other, and their environment. Ants do not think. They do not consciously plan. They have no overall concept of what an ant colony should look like. Instead, what they do is to follow simple rules of behaviour.

We can also learn from the way in which ants search for food. They cannot plan their route, nor do they ever really know where they are. To make sure that they can find their way home again, they leave a pheromone trail behind them. Essentially, they search at random until they stumble upon something. They then report back to the nest. If, in the course of searching, an ant stumbles on a pheromone trail, then it follows that trail because it may lead to food that some other ant has discovered.

Like most houses on the outskirts of cities, during summer we are often plagued by small ants breaking into the house. If you watch where they go, you find that they often follow very convoluted paths—down the wall, across the sink, down to the floor, along the skirting board, and eventually into the cupboard. If you interfere by wiping their trail, they quickly find it again by wandering around at random in the general area until they manage to cross the path again. Once one ant finds the link, others follow and within a minute or two the trail is re-established. Over time the kink where you cut the trail is smoothed out, until it disappears.

One secret of their success at searching is that the ant colony sends out not just one scout, but thousands. But thousands of scouts will not be any more efficient if most of them are searching fruitlessly, and only one or two find the pile of crumbs that you accidentally dropped on the floor. Pheromone trails make it possible for individual ants to change their tactics and home in on rich resources.

Computer scientists are attacking complex problems by emulating the ants. Internet search engines, for instance, use agents to travel around the Web looking for new resources to add to their index. Their paths are essentially random, since they follow up whatever links happen to lie within the documents they explore.

One idea is to imitate the idea of pheromone trails on the Internet. Duplicated effort is a waste of time, so if one of these virtual ants finds something that can be useful to another ant, that information has to be transmitted from one to the other. There are several competing ideas for implementing pheromone-like trails. One is to provide a log on a server where incoming ants can leave messages. They can file a report of their visit, including where they came from, what they were looking for and whether they found anything useful. Newly arrived agents could then save time by first looking through the log for advice from past visitors.

Of course, searches for online resources differ from the kind of searching that ants do in one fundamental way—the food that ants track down is a finite resource. Once it is eaten, it is gone, so the ants concentrate on the richest resources, exploiting them as much as possible before other bugs join in. Online data, however, do not normally disappear when they are accessed! Thousands of users can look at the same resource again and again, so the imperative is not rapid exploitation, but efficient retrieval.

Agents in the electronic marketplace

You want to buy a boat before the start of summer. You find a terrific deal, but only if you can raise the cash quickly. You need a loan so you log into your bank's Website. There you 'talk' to the bank's online loans broker by supplying the relevant details when prompted. At the end, you click on the 'OK' button and the broker comes back with the message 'Congratulations, your loan is approved'. You have been talking to an agent.

The use of agent technology by the business community is spreading rapidly. The trend has been driven by the rise of electronic or 'e'-commerce and accelerated with the arrival of a viable public interface in the form of the World Wide Web. The advantages are the same as any form of automation. Agents are best suited to handling routine, repetitive tasks that require only limited ranges of knowledge.

The commercial world discovered the advantages of the World Wide Web almost as soon as it was launched. Commercial sites first started appearing on the Web in significant numbers during 1994. At first, most of these sites were service providers. During 1995, the number of commercial sites began to climb rapidly and, by 1996, commercial interests had begun to dominate the Web. By the turn of the millennium, e-commerce had become big business. Internet enterprises were among the top grossing companies. For a time, investments in stocks of Internet companies boomed in a way rarely seen before.

Software agents play a big part in e-commerce. They go by many names, such as bots or brokers, and operate in many contexts. The public face of agents consists of the services that people can access online. Some of these agents reflect the traditional activities of the marketplace. For instance, *buyer agents* perform tasks on behalf of customers, such as locating and comparing products and sources, and even making purchases. *Seller agents* handle sales for vendors, track changes in demand, watch competitors and undertake promotion. *Broker agents* act as intermediaries to bring buyers and sellers together. In a sense, they are the e-commerce equivalents of stockbrokers, or even a marketplace.

As with other applications, there are dangers, as well as advantages, in e-commerce agents. One is the potential for electronic snooping, which is discussed below. Another danger arises from 'pathological' interactions between agents.

On Monday, 19 October 1987, the US stockmarket crashed. On what became known as 'Black Monday', the Dow Jones Industrial Average fell 508 points, the largest drop in a single day in the market's history. At one point it looked as if the entire US financial system might collapse. Although the causes are still being debated, electronic trading was probably a significant contributor.[4] Ten years later, on 27 October 1997, an even larger fall of 554 points occurred. However, in this later incident, the market immediately bounced back up, recovering 337 points the following day.

Stock market traders use software that buys and sells according to trends. Traders use electronic tools because they allow them to respond very quickly, to subtle shifts in the market. Electronic trading has transformed the stock market. In many cities, frenzied pits filled with chalkdust and shouting have been replaced by stony silence and electronic bulletin boards, but a more dangerous effect has been to make the market far more volatile. The high speed of electronic trading leads to feedback in which up and down trends quickly run away before they can be checked.

Big brother is watching

In George Orwell's novel *Nineteen Eighty-four*, people lived in a society that was almost totally controlled. Wherever you went, whatever you did, you were on camera. Your every word, your every action was monitored. Orwell's book was an argument against the evils of communism and other totalitarian systems. It was also a warning of what could happen in any society if there were no privacy and too much information in the hands of too few.

Modern society does not yet have cameras on every street corner, and in every building.[5] We are not being watched every minute of the day. However, in the age of electronic information, we are constantly leaving clues behind that give us away.

It is a common movie scenario. The hero walks in for a meeting with another character, who picks up a dossier off the desk and proceeds to read out a detailed summary of everything the hero has ever done. Forget about 1984, Big Brother is here. Governments can glean virtually any information they want about you. But that is only the start of it. As Thomas Friedman points out in his book *The Lexus and the Olive Tree*, privacy invasion today does not just mean Big Brother. Anyone with enough money, and a mind to, can find out just about anything they want about you. Some companies can and do check up on clients, competitors, even their own staff.

As any good detective story will show you, you can learn a lot from the clues that get left behind. Palaeontologists study ancient life forms by seeking out clues in the form of fossils. These fossil clues include many different kinds of evidence, from bone that has turned into rock to pollen grains trapped in lake sediments, to isotope levels under ice caps. Archaeologists can learn a lot about ancient civilisations from the rubbish that they left behind. Likewise, modern sociologists have turned to rubbish bins to gather clues about today's lifestyles. These 'garbologists' can deduce a surprising amount about what people do and even about what they think and feel just from studying the junk they throw away.

In the same way, you can learn a lot about people from the electronic junk they leave behind them. Not so many years ago, if you opened someone's purse or wallet, you would find notes and coins, and perhaps a driver's licence. Today you will find plastic—bank cards, credit cards, discount shopping cards, phone cards, travel cards, petrol cards, club memberships. Even the driver's licence is now plastic.

Each of these pieces of plastic has a magnetic strip embedded in it. On those strips are all the details of your bank account, credit limit, or whatever sort of information the card deals with. When you use your card in a supermarket, the cashier swipes it and the details get electronically recorded. The result is that you leave an electronic trail of data behind you wherever you go.

The question is, what happens to all that data? Now it may not matter much if some company has it on their files that you purchased soap at such and such a shop on such and such a day. But that data is shared. If you used your credit card to buy the soap, then the shop will have to pass the details of the purchase on to the credit card company (usually at the point of sale). This means that organisations such as credit card companies and banks can create a profile of your spending activities.

A growing concern is that a lot of personal information is becoming much more widely available. The proliferation of personal information on the Internet has made it easier than ever to gather details about people. About the time that the Internet started to become really popular, I upset a reporter who interviewed me about privacy on the Internet. He had no idea that there was any information about him online and was shocked to discover that a few minutes' search turned up not only his home address and telephone number, but also his professional history and even a photograph. More time and determination (and parting with a bit of cash) would have yielded a lot more.

Some people, especially those in professions, have access to a lot of public information online. Other people have access to very little. In today's world, whatever you do, your actions leave an electronic trail in public places. If you own a phone (and do not pay for a silent number), you will be listed in the phone book. And the phone book in many countries is now online. Governments, and certain corporations, compile substantial databases about you and your activities. As we saw above, organisations such as banks and credit card companies can put together a detailed sketch of your spending activities. More and more of this information is appearing in online resources.

Apparently innocent pieces of data can be revealing when they are combined. If your credit card spending suddenly increases from hundreds to tens of thousands of dollars, then it is fair to infer that something unusual is going on. By combining data on credit card spending and bank statements it would be possible to build a picture of an individual's finances. Perhaps the most common use of all this data at present is to detect market trends and patterns, not to monitor individuals. But more and more companies are trying to target their advertising at the individual.

Have you ever had this happen to you: I phoned up a company to enquire about something and the anonymous person at the other end greeted me by name. 'Good morning Mr Green, how are you today?' The personalised greeting is aimed at putting callers at ease, at providing a more friendly service. Instead, I find it slightly disturbing that some big corporation already knows so much about me! What happens is that as soon as my call was received, my phone number was recorded and fed into a database. By the time the operator answered my call, my complete file was displayed on a monitor. Not only that, but the details of my current call would most likely be added to the file.

Much the same thing happens on the Internet, but unlike your phone number, the address that you link from is not a reliable guide to who you are. For instance, many systems assign random Internet addresses. So your email address is likely to be different every time you connect. One way that sites overcome this problem is by using *cookies*.

A cookie is a short piece of data that passes between a Web server and a user's browser. Its main role is to help the Web server to maintain continuity in its interactions with users. Another aim is to provide more user-friendly services by keeping track of a user's previous activity.

The key element of a cookie is a code that identifies the user. This code allows the Web server to look up relevant background details of the user's previous activity. However, the code could be anything at all. For instance, it could encode details of the type of operation that you are currently undertaking.

On the Internet, Big Brother really is watching everything we do. Most people do not realise it, but every time they access a Website, the server makes a record of that transaction. Here are some typical entries that might appear in a Website's access log.

```
nurk.college.edu [12:16:20] /penguins/ HTTP/1.1
joe3.megazzz.com [12:16:33] /bigsnooze/mattress.html HTTP/1.0
drone.engine.com [12:16:41] /robots.txt HTTP/1.0
joe3.megazzz.com [12:17:54] /bigsnooze/spring.html HTTP/1.0
1122.777.999.888 [12:17:56] /courses/ HTTP/1.0
joe3.megazzz.com [12:18:51] /bigsnooze/prices.html HTTP/1.0
host.nobody2.net [12:19:02] /cgi-bin/search?data HTTP/1.1
joe3.megazzz.com [12:20:21] /bigsnooze/deals.html HTTP/1.0
```

Each record includes the user's address, time of access and the item accessed. Records of this kind would not pin the access down to any individual, but a pattern of activity from a particular domain, combined with other information, could provide strong hints as to the activity and interests of particular individuals. So the entry from drone.engine.com appears to be a search engine agent automatically scanning certain entries. The pattern of accesses from joe3.megazzz.com suggests that someone from a bedding company is checking up on a competitor.

The error log of a server is often more revealing still. It records requests that failed for one reason or another. Most often this happens when you try to download an item that's moved or been deleted. However, if someone comes snooping for hidden information on your Website, the visit usually stands out a mile on the error log.

In 2000, the British government revealed plans to install a system which would monitor all email transactions made through Internet service providers.[6] The aim of the scheme was to curb the increasing use of the Internet for criminal activity, but it would also enable the government to eavesdrop on anyone's personal emails. In this case, the proposal was abandoned because it was too easy to thwart the monitor.

At the time of writing, the balance of state monitoring versus personal privacy is arousing much concern. The threat of global terrorism is forcing many western governments to seek greater powers to guard against the acts of terrorists. Although motivated by concern for public and community safety, an increase in government powers to monitor individuals and their activities is itself a potential threat to civil liberties.

chapter 8

COMPUTING AND NATURE

Whence is it that nature does nothing in vain; and whence arises all that order and beauty which we see in the world?

Sir Isaac Newton, *Opticks*, 1704

The history of science is a story of humankind gradually learning humility in the face of nature. Over the centuries, one discovery after another has moved humans from their position at the centre of the universe, made in the image of God, to that of a recently evolved hominid species. Along the way, we have learned to wonder at the remarkable things that animals and plants can do. How does a spider build such a beautifully symmetric web? How do some birds and fish manage to navigate thousands of miles to find their breeding grounds? As scientists struggle to cope with problems of increasing complexity, they are finding that nature has evolved many of the solutions they seek.

Every age has had its own way of looking at the world. During the Industrial Revolution, society was obsessed with machinery. The world was viewed as a great machine. Nineteenth-century science provided mechanical explanations for many natural phenomena: the solar system ran like clockwork; the body was like a factory, with the heart its engine, the brain its control room, and so forth.

By the end of the twentieth century, computers and information were the new obsessions. In this Information Revolution, even nature is seen as a form of computation. Modern science tends to treat the world, and everything in it, as 'natural computing'.

A new paradigm

Life is a form of computation. This compelling analogy is one of the most potent and productive ideas in all of theoretical biology. One motivation behind it is that many natural processes do behave like computation. The genetic code is often compared to a computer program because it contains the code for building new organisms. Ribosomes behave like tape drives that 'read' RNA and output protein. We can represent animals as 'agents' that behave according to rules, much like programmed robots. Sensory perception involves processing data input and animal communication involves the transfer of information from one animal to another. Finally, the brain performs a variety of information processing tasks. While it is risky to take such analogies too literally, ideas from computing can explain certain aspects of many natural phenomena.

The links between computing and the natural sciences, especially biology, are now very strong. Increasingly, computational models (especially simulations) are used to study biological and physical processes. But the paradigm of natural computation works in both directions. Biological ideas have also crept into computing. Active fields of study include such topics as cellular automata, genetic algorithms, neural networks, evolutionary programming and artificial life.

One reason for this synergy between biology and computing is that scientists have realised that living systems evolve practical ways to solve all manner of complex problems. For instance, all animals have to solve the problem of finding food—the methods they use vary according to their nature and the kinds of food they need to find. Some animals, such as hawks or sharks, use exquisitely honed physiological senses to detect and track down prey. Others, such as bees and ants, use parallel processing. They send out hundreds of individuals, who coordinate their hunting by signalling their fellows for help when they find something.

Animals and plants also need to solve problems of optimisation. Which strategy is better for a growing plant—to grow fast and produce lots of seeds? Or to develop adaptations to help it survive during drought? Living things also have to partition time. The longer a bird spends searching for food, the better it can feed its chicks, but the longer it is away from the nest, the greater the danger to those chicks.

Ultimately, the success of plants and animals at solving such problems boils down to how well their adaptations and behaviour match their environment. The speed of a cheetah would be out of place in a dense forest where prey are more likely to hide than run. Likewise, we humans learn to interact with the world by developing internal models of our environment. Many psychological disorders amount to a mismatch between those models and reality, such as a trauma victim finding innocuous everyday events threatening.

Many areas of advanced computing are almost indistinguishable from biology. *Artificial Neural Networks* (ANNs), for instance, derive inspiration from the workings of the brain. They consist of nodes ('neurons') that pass data from one to another, performing simple operations as they go. In a typical, multi-layered network, the bottom layer of neurons accepts inputs from the outside world and passes it up to neurons in the next layer. Data are then passed from layer to layer until they emerge from the neurons in the 'output layer' at the top. Users can calibrate the network by presenting it with training examples.

Machine learning, which includes ANNs, draws on studies of the ways that humans and animals learn. In another model known as a *decision tree* (see Chapter 7), the system organises rules into a hierarchy that enables a computer program (or a human interacting with it) to solve a complex problem in stages. We humans use decision trees all the time without being aware of it. For instance, do you want a drink? Tea or coffee? White or black? Milk or cream? Each question is like a branch, with the two possible answers leading on to further questions, thus narrowing the range of possible drinks.

Processing in swarms

For computer scientists and engineers, one motivation for studying natural computation is to learn how to design supercomputers and grids. As we saw in Chapter 5, large-scale computing relies increasingly on having many processors to do the job. These processors may be concentrated inside a single box, clustered in arrays, or a grid of widely dispersed computers. They can achieve spectacular results, but how should they be organised?

For problems in which the processors can act independently, their organisation is simple. A central coordinating node passes out tasks to each processor and gathers the results when they finish. But what if the nodes need to 'talk' to each other while they work? For example, in a model of a traffic jam, an epidemic, or a galaxy, the elements which represent cars, people, or stars, need to interact with each other constantly. Also, the way in which the nodes are organised needs to change according to the problem. In old-fashioned telephone systems, a switchboard operator used to connect callers by inserting a cord into a plugboard. In the same way, we need to be able to connect lots of processors together. Different problems need different wiring patterns. For instance, in a simulation model of global weather patterns, each processor might correspond to an area of the earth's surface or to a cell of air above that surface. As they mimic the flow of heat and the movement of air through the system, adjacent cells in this kind of model need to carry on a constant dialogue. A model of the global economy would need a completely different kind of organisation. Each cell might correspond to a company (or country) and the cells would need to mimic financial transactions and the flow of goods.

It is not surprising, therefore, that scientists have looked at natural systems to find new ways of organising large-scale computer processing. From living systems, they can also learn how large-scale behaviour emerges out of the local actions and interactions of the individual cells.

Swarm intelligence

Many ants, all obeying simple rules, create the order that we see in an ant colony. This is an example of what has come to be known as *swarm intelligence*: behaviour or design that emerges out of simple responses by many individuals. Understanding how this happens is important in designing systems of components that have to coordinate their behaviour to achieve a desired result. Knowledge of the way order emerges in an ant colony, for instance, has been applied to create the so-called *ant sort algorithm*, which is used in contexts where items need to be sorted constantly, without any knowledge of the overall best plan.

The most familiar example of swarm intelligence is the human brain. Memory, perception and thought all arise out of the nett actions of billions of individual neurons. As we saw earlier, artificial neural networks (ANNs) try to mimic this idea. Signals from the outside world enter via an *input layer* of neurons. These pass the signal through a series of hidden layers, until the result emerges from an *output layer*. Each neuron modifies the signal in some simple way. It might, for instance, convert the inputs by plugging them into a polynomial, or some other simple function. Also, the network can learn by modifying the strength of the connections between neurons in different layers. For instance, suppose that it applies a choice function such as:

> Toss a coin 10 times.
> If there are more than 5 heads,
> then output input 1,
> otherwise output input 2.

If the resulting output is correct every time it chooses input 1, but wrong whenever it chooses output 2, then the number of heads required to choose input 1 will be gradually relaxed, until any number of heads will do.

Swarm intelligence occurs in many other contexts. The term itself refers to the ways in which bees, birds, fish and many other animals combine to act as a swarm, a flock or a school (see Figure 8.1). In most cases, there is no central controller. For example, the basic flocking behaviour of birds arises from the following three simple rules:

- Keep as close as possible to the centre of the group.
- Avoid crowding too close.
- Keep to the same heading as the rest of the group.

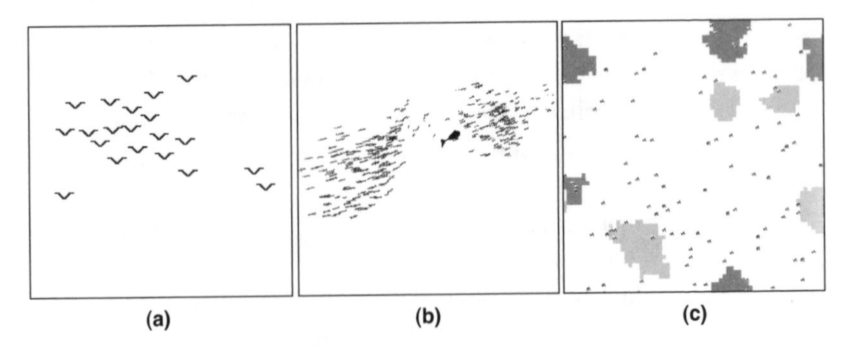

Figure 8.1 Models of emerging group structure
(a) a flock of birds; (b) a school of fish; (c) an ant colony.

Robot insects

Mention the word 'robot' and most people think of a mechanical man. The image may look something like Robby the Robot from the film *Forbidden Planet*, or R2D2 in *Star Wars*, Commander Data in *Star Trek*, or perhaps Robyn Williams's character from *Bicentennial Man*. Whatever particular fictional image the word robot conjures up for you, the truth is that researchers still have not produced an intelligent, human-like machine.

At present, you are most likely to see a robot on a factory assembly floor. In contrast to the movie image of a metal man, a real-life robot is likely to be nothing more than a mechanical arm attaching panels to the side of a car. Robots are now commonplace in industry but, more often than not, they are based on a relatively simple technology. A robot on a production line can get away with being dumb—having a device that will perform the same operation reliably and precisely, time after time, can be a great advantage for productivity. It also ensures that the quality of the finished product is of a consistently high standard.

Such robots have some crucial limitations, however, which restrict their potential applications. For one thing, they cannot adapt to changing situations. They rely on the objects that they manipulate being presented in a consistent way. If a panel turns up back to front, for instance, they will to try to weld it in backwards. Not only can they not interrupt their routine to turn the panel around, they don't even 'know' that it's the wrong way round. Human operators, in contrast, would instantly recognise and correct problems of this kind.

Another limitation is that the robots are centrally controlled. A single, large computer acts as their 'brain', controlling every aspect of their operation. This top-down, centralised control is not only clumsy, but also slow and awkward. A common phenomenon with robots of this kind is that if they encounter an obstacle, they will stop what they are supposed to be doing while their central brain tries to analyse the scene and decide what to do.

We saw earlier that the organisation of an ant colony emerges from the behaviour of lots of ants obeying simple heuristics, or rules of thumb. In the same way, a new generation of adaptable robots is now being built in which the robot's behaviour emerges out of the joint activity and interactions of lots of local controllers for its legs, its senses, and so on.

In the 1980s, Rodney Brooks and others succeeded in showing that neither central processing, nor intelligence, was needed for the operation of a successful robot.[1] In Brooks's robot insects, each leg is individually controlled and responds separately according to a set of simple rules. For instance, one local rule for leg function might be:

> IF leg is pointing back,
> THEN lift leg and move it forward.

Interacting feedback loops coordinate the legs to control walking and other fundamental tasks.

Perhaps the most dramatic indication of the superiority of this new approach to robotics is the shrinking size of the new machines. Early mobile robots were usually large and clumsy, but distributing control has made it possible to simplify and shrink machines until they are the size of large insects. In addition, smaller and simpler machines are also easier and cheaper to manufacture.

There are many possible uses for tiny robots. In the foreseeable future, for instance, they could become a common feature in homes and offices, wandering around at night cleaning up dirt, polishing the floor and putting things away. But there are other, less mundane, uses for these machines. Small, self-controlled robots can explore places where it is difficult, dangerous or impossible for people to go. But perhaps their greatest potential lies in acting in unison. Like real insects, they could search, sort and organise.

Another kind of flexibility lies in their ability to cope with different terrains and climatic conditions. Most animals are limited by their physical characteristics to particular environments. But why should machines be limited in the same way? Scientists are experimenting with robots that can change their shape to adapt to different surfaces. For instance, if the robot's body consists of many flexibly linked modules, then different conformations of those modules can enable it to wriggle like a snake, roll like a car, loop like a worm, or rise up and walk like an insect.

Self-replication

One of the hallmarks of a robot is that it should act independently of outside supervision. If we take this idea to its logical conclusion, then ideally robots should not only be able to act alone, they should also be able to manufacture themselves. That is, they should be self-replicating.

The challenge of creating a self-replicating robot is identical to that of devising new methods of manufacturing by which a single, universal process can produce and assemble many different kinds of objects. The ultimate answer to the dream of self-replicating machines may lie in developing completely new methods of manufacturing. Many people see biological methods as the solution: devices would literally be 'grown' from organic components. There is a real possibility that advances in our understanding of genetic control over developmental processes (see Chapter 9), may eventually make such technology possible.

In the Middle Ages, theologians introduced the idea of the *homunculus* to explain reproduction. At birth every woman carried homunculi in her womb. These were miniature copies of humans, from which babies would grow. However, this idea posed a paradox. Each homunculus would need to hold even smaller homunculi and so on for every future generation.

How is it possible for a system to include all the details necessary to reproduce itself? Surely, like the homunculus, the information required amounts to a complete copy of the individual, including the description of itself? Like a pair of mirrors reflecting each other, the problem seems to spiral off into infinity. Yet, somehow, living organisms demonstrably resolve the paradox.

The solution seems to be that organisms do not need to encode a complete description of themselves, just the recipe for making a copy. Most of the detail is supplied by the context in which an embryo develops. Early investigators into self-replication, such as John von Neumann and Lionel Penrose, demonstrated that relatively simple systems could replicate themselves without the need for complex biological machinery.

The first challenge is simply to reproduce a given design. An important step was to demonstrate the process of self-replication in simulated organisms. Tom Ray's model Tierra did exactly this.[2] In Tierra, the 'organisms' are programs that compete for space in a finite universe of computer memory. Each organism includes code that allows it to make copies of itself. Organisms can mutate and evolve, but if they lose the ability to replicate, then they leave no descendants. Although the model started with a given routine for copying programs, several alternative strategies emerged, including a more compact method than the original. Perhaps the most surprising outcome was that parasites appeared spontaneously in the model. These organisms possessed no code of their own for reproduction but they developed the ability to use the code for replication in other organisms.

Climbing virtual mountains

The problems associated with writing programs to solve large, complex problems are often so difficult that scientists have to look for alternatives. Instead of writing a program explicitly, perhaps you could train it by using examples? Could you grow it by combining smaller elements, each of which do part of the job? Perhaps you could even evolve it by working with a population of 'solutions' and breeding the most successful ones?

One of the revolutions that computers have brought about is that we are no longer restricted to traditional methods of solving problems. Here 'traditional' means writing down a problem as a set of equations and then proceeding to solve the equations to find a solution. Normally we can reduce a problem to a set of *parameters*—numbers that we need to find. For instance, if a shopkeeper wants to achieve the maximum profit on a new line of shoes, then he needs to set the price at just the right level. If the price is too high, then few people will want to buy them. However, if the price is too low, then the profit on each shoe sold will be too low. He has to strike a balance where the total profit is as great as possible. If he knows how the volume of sales varies with price, then it is a simple matter for him to solve the equations and work out the optimum price for the shoes.

Unfortunately, many problems are too hard to solve this way. When lots of factors influence the outcome and when those factors interact with each other, the equations often become unsolvable. The advantage of computers is that you can try out hundreds or even thousands of possible solutions to a problem in no time at all. Given this, the ancient approach of 'trial and error' becomes a viable way of solving problems.

The simplest form of trial and error is known as *hill climbing*. The analogy is that of a mountaineer trying to find the top of a mountain during a snowstorm. Although he cannot see the top, he can reach it eventually by walking uphill. In the same way, we can often reach the optimum solution by systematically tuning the values of the problem's parameters. Unfortunately, not all problems can be solved by this simple approach. The mountaineering analogy helps us to see why: walking uphill will get you to the top of the mountain if there are no foothills, and if you are on the mountainside in the first place. If you are on a foothill, then you will reach the top of the foothill, not the mountain.

In the same way, we can imagine the solutions to a problem as a sort of landscape. Changing the values of a parameter is like going north–south or east–west. The elevation corresponds to how good the solution is. This analogy leads to the notion of a *solution landscape*. In reality, the 'landscape' may actually have more than three dimensions—many more. Nevertheless, it makes a convenient way to visualise what is

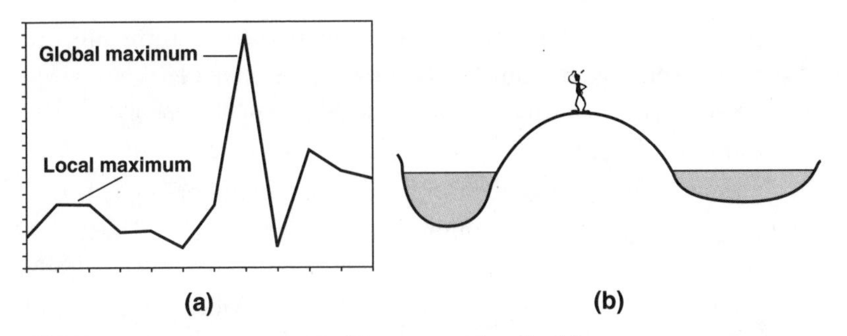

Figure 8.2 Optimisation is like mountain climbing
(a) How do you find the top of the highest peak without becoming trapped in the foothills? (b) The great deluge algorithm. The valleys fill with water forcing the mountain climber to seek higher ground.

involved in solving problems. To begin with, the landscape may have many mountains and hills. This means that many different solutions are *local optima* and, for this reason, the hill-climbing method is known as a form of *local search*. If, however, we want to find the *global optimum*—the highest mountain, the best solution—then we need to conduct a *global search*. How does our mountaineer find all the hills? Wandering around at random is slow and wasteful. If he has a helicopter, then he can fly up above the clouds, descend at different spots and check the local elevation. If an area looks promising, then he can wander around a bit to check whether there are any nearby hills. Eventually, when he runs short of fuel, he can fly to the highest peak that he has found.

Many optimisation algorithms are variations on this approach. Any algorithm must involve some combination of global and local searching. In the *Great Deluge*, for example, our explorer is trying to escape from a biblical flood. The rains have come and the landscape is gradually flooding. At first, he can wander almost anywhere. By following ridges, he can skirt the rising waters and travel from one peak to another. Eventually, however, the flood turns the peaks into islands. He becomes trapped on one of the mountains and has to work his way to the top as the dry land gradually shrinks away.

Other algorithms achieve the same thing in different, yet similar, ways. *Simulated Annealing*, for instance, draws on an analogy with metallurgy.

Instead of rising water, it is falling temperature that controls the process. Just as heating during annealing allows atoms in a crystal lattice to settle into their optimal arrangement, so the algorithm jiggles its way to the optimal solution to a problem.

In these and other similar algorithms, the underlying principle is that the tallest peak—the global optimum—is likely to occupy a larger area than other hills in the region. This means that our mountaineer is likely to spend more time on the largest mountain than anywhere else, and will probably be trapped on that mountaintop at the end of the process. However, it is important to realise that these methods do not guarantee that we will find the very best solution. Are any of these methods any better than a random search? They are all variations of trial and error. The question is whether they get you to the best solution faster, or more reliably, than a random search?

Evolving answers

Nature is blind. Nature is dumb. Nature does not solve problems; it evolves answers. In *evolutionary computation* ('EC'), computer scientists have developed many problem-solving methods by imitating biological evolution.[3] Among the earliest, and most widely used, of these methods are *genetic algorithms*, which were popularised by John Holland.[4]

To understand how a genetic algorithm (GA) works, let us suppose that a car manufacturer wants to design a car that minimises fuel consumption, while still maintaining acceptably high levels of power, comfort, safety, and so on. Drawing on long experience, the designers have developed complex numerical models that tell them exactly how a car will perform under any combination of conditions. The trouble is that performance does not change in simple fashion, so simple methods cannot solve the problem. To use a GA you express every assumption, parameter and feature that affects the behaviour of the model as 'genes'. You can then represent any potential solution to the car design as some combination of genes, or 'genotype'. Instead of just one model, you form an entire population of potential solutions, each with its own genotype. Now comes the fun part. You breed these solutions with one another to

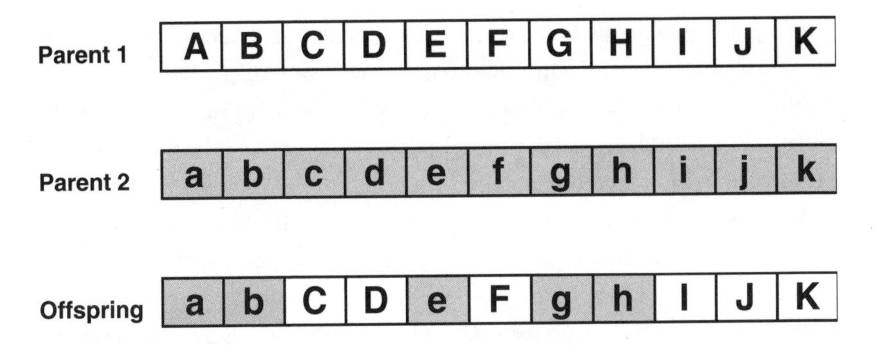

Figure 8.3 Operation of a genetic algorithm
Each genotype (a . . . k) consists of values for parameter and assumptions. To form the genotype of each offspring, genes are selected from the two parents ('crossover').

form a new generation of models. These new models are the children of the models that you started with.

Next, you cull solutions that perform too poorly. The survivors become the parents for the next generation. You continue this process for a series of generations. As generation succeeds generation, the algorithm breeds better and better solutions. Eventually, it gives you solutions that meet your needs.

At first sight, the GA may seem like an extravagant way to tackle a problem. Instead of a solution, we take an entire population of solutions. To use our explorer analogy again, it is as though we scatter lots of explorers all over our landscape. But here the analogy breaks down, because the solutions at higher elevations breed with one another to produce offspring. These offspring form a new generation, which then breeds in the same way. In each generation the solutions at the lowest elevations miss out.

Computer scientists have introduced many variants of the basic GA. For example, the *generation* approach replaces an entire generation all at once. In contrast, a *steady-state* GA replaces individuals one at a time. Many of these introduced variations are attempts to improve the performance of the algorithm or to adapt it to different kinds of problems.

For many problems, it is not so easy to reduce the issue to a simple set of genes. For instance, suppose that we want to devise an automatic method to process digital photos. The aim might be to make the picture clearer. The trouble is that not only are there many possible operations to perform on the image (e.g., sharpening, colour balancing), but also the order in which they are performed is important. The problem is that you cannot cross two sequences of operations directly. For instance, suppose that we make a list of the image-processing operations above and number them 1 to 5. Then a sequence such as (1 2 3 4 5) would provide a recipe for processing the image. But if we tried to cross two sequences, such as (1 2 3 4 5) and (5 4 3 2 1), we would get sequences such as (5 4 3 4 5) in which some steps are repeated, and others omitted.

What we have to do in this case is make the genes work at a deeper level. Instead of representing the model directly, they represent steps involved in building it. We can build sequences from a string, such as (1 2 1 1), where the number '2' means 'take the second unused number from the list'. So this string would yield a sequence (1 3 2 4 5). We need only four numbers because the last item in the sequence will be whatever remains at the end. So a code string (3 1 2 1) would lead to the sequence (3 1 4 2 5).

The point is that the encoded strings provide a set of genes on which we can use a GA. For instance, if we cross the strings (1 2 1 1) and (3 2 2 1) then we might obtain strings such as (1 2 1 1) or (3 2 1 1). In each case, the result will be a viable string from which we can build a processing sequence.

The wellspring of creation

Nature exploits chaos as a source of novelty.

Medical researcher Walter Freeman[5]

Perhaps the most exciting applications of evolutionary computing are in art and design. In his book *The Blind Watchmaker*, Richard Dawkins introduced the idea of a *biomorph*.[6] A biomorph is an artificial creature that can evolve its shape in response to selection by a human player. The

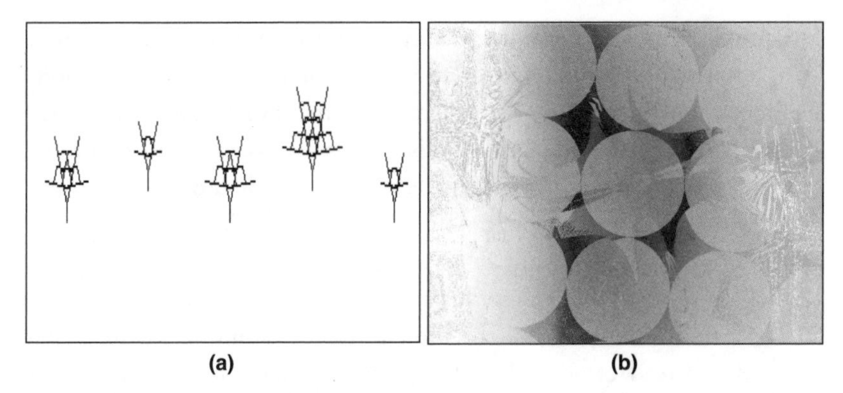

Figure 8.4 Examples of evolutionary design
(a) Biomorphs (b) Cambrian Art.

aim was to demonstrate how selection can shape an organism over time. In each generation of Dawkins' biomorph game, the computer takes the existing creature and produces several variants. The player then selects which variants will become the parent for the next generation. By carefully choosing attributes, a player can create creatures of almost any design.

The idea of applying this approach to art and design did not take long to catch on. The Cambrian Art Web site, for instance, allowed users to create a colourful range of abstract pictures by applying Dawkins' idea.[7] There have also been commercial applications. Architects and car manufacturers use similar methods to come up with novel designs.

Does nature know best?

As we have seen, one attraction of biologically inspired approaches is that nature has thrown up many ways of solving complex problems. However, there are many opponents of biocomputing. This heated debate centres on methods of finding optima. Opponents argue that analytic methods are more reliable and more efficient than GAs and other biologically inspired methods.

One source of contention is that people often use a sledgehammer to crack a walnut. GAs are popular because they can be applied to almost any problem, but this advantage leads to indiscriminate use. Uncritical users tend to apply them to everything, even where simpler methods would be more effective. Also, there is no free lunch: every advantage comes at a price. Traditional methods of optimisation are designed to be efficient and reliable—their price is that they are limited by their assumptions to a narrow range of applications.[8] GAs can be applied to any problem, but the price is that for any given problem, there will always be some specialised method that will perform better.

Another issue arises from the limitations of traditional optimisation methods. Instead of addressing the problems they *should* solve, people tend to address problems they *can* solve. They make simplifying assumptions to narrow the scope. Companies, for example, are always under pressure to maximise annual profits. Taken in its narrowest sense, finding a way to maximise next year's profit may be a relatively simple problem. But if it means laying off staff, selling your assets and taking out huge loans at exorbitant interest rates, it could be suicide in the long run. In cases like this, we really have multiple objectives. When different objectives interact in complex ways, traditional methods may not work.

A final point is that GAs are not guaranteed to find the optimum. Traditional optimisation methods aim to find the very best solution possible. However, that is not what happens in nature. An antelope, for instance, does not need to run as fast as is physically possible. It is enough that it can outrun potential predators. A tree does not need to catch every photon of light that falls upon it, just enough to allow it to grow. In other words, living things find adequate solutions, not the global optimum. For living things, survival does not mean gorging yourself today and letting tomorrow look after itself. It means surviving and reproducing, generation after generation, in the face of all possible contingencies. GAs are really answering a different question.

chapter 9

PANDORA'S BOX

Order is Heaven's first law.
Alexander Pope

Following its discovery, the double helix model of DNA was instantly hailed as the secret of life.[1] Its significance was that it immediately suggested ways in which the genes worked. Two chains formed the helix and these would split during reproduction, thus making two copies. The sequences of bases that lay along each of the helix chains were even more exciting. By identifying them, scientists had at last found the physical elements that made up the genetic code.

Fifty years later, the human genome has been mapped and scientists have learned how to cut and splice genes and to create natural products on demand. By the turn of the millennium, genetic information was like a rocket taking off. Bioinformation was fast becoming an important currency in the global economy.[2]

Life's program

The analogy of life as natural computation is nowhere more evident than in the genetic code. The idea of the genome as a computer program is very compelling. In many respects, the details of the genetic code serve to strengthen the analogy, although there are important differences. To begin we need to understand what genetic information consists of.

The sequence of bases along the inside of the DNA double helix forms the genetic code. This is possible because of a remarkable property of the four bases involved. What happens is that two carbon-phosphate chains form the outside of the DNA double helix. The bases attached to the two chains hold the structure together by pairing with each other across the centre of the helix. The crucial property is that when the bases Adenine (A) and Thymine (T) link together via hydrogen bonds, they form a structure that is nearly identical in size and shape to the structure formed by Guanine (G) and Cytosine (C). So A-T is structurally equivalent to C-G, and to G-C, and to T-A. This means that at as far as the helix is concerned, it is immaterial which pair of bases occurs at any point. Nor does it matter which way around they are oriented. The physical structure has no bearing on them, which means that the bases are free to order themselves according to biological factors, namely the genetic code.

If written out, the human genome would be a string consisting of the four letters mentioned above: A, C, G and T. If it were written out as a single string, it would be about 3 300 000 000 symbols in length. If it were printed out as books the size of this one, it would fill a library of over 50 000 books. In terms of computer disk storage, that would be over 3 gigabytes of data (assuming one byte per base pair). If you sat down and tried to type it out, at the rate of one symbol per second, and supposing that you could work without any breaks or rests at all, then the job would take you nearly 105 years.

These figures are even more amazing when you consider that every cell in the body is able to complete this task in a matter of hours during cell division. And there are about 75 000 000 000 000 cells in the body each containing a complete copy of the human genome. Furthermore, humanity is just one of millions of species on the planet, each with its own distinctive genome. The human genome is neither the largest, nor the smallest. The total amount of genomic information in the world is astronomical. If we suppose that each genome is about a billion base pairs in size, then the total data for all species exceeds 1 000 000 000 000 000 base pairs. In computing terms, that is at least a petabyte of data.

These figures highlight the crucial role of information technology in biotechnology. Without computers to store and interpret the data, it would be impossible to make any headway at all towards understanding the genetic code.

Following the discovery of the genetic code, the next important step was to learn what it meant. The mechanism of turning the DNA code into proteins involves intermediate steps. First, the DNA strand is copied to produce a new chain called *messenger RNA*.[3] The analogy of RNA with a data tape is very strong. Just as a computer reads a tape, ribosomes read RNA strands.[4] The ribosomes are effectively small biochemical factories that translate the sequence of bases into proteins.

A protein is just a polypeptide chain, consisting of a string of amino acids linked together. To build a protein, a ribosome creates amino acids, one after another, according to the genetic code. In the genetic code, each *codon* or sequence of three bases codes for a single amino acid. The codes are listed in Table 9.1. In the following segment of DNA, the top string gives the sequence of bases and the bottom string gives the corresponding sequence of amino acids, as given by the table.

Bases

```
GAGTATCGCTTCCATGACGCAGAAGTTAACACTTTCG-
GATATTTCTGA
```

Amino Acids

```
Glu Tyr Arg Phe His Asp Ala Gln Val Asn Thr
Phe Gly Tyr Phe STOP
```

The triplet AAA, for instance, codes for Lysine. Notice that most of the amino acids appear in the table several times. The triplet AAG also codes for Lysine.

An important property of the genetic code is the notion of a *reading frame*. This is a segment of the DNA sequence that can be read and translated into the amino acids that make up proteins. The start of a reading frame is indicated by the start codon ATG, which also codes for Methionine. The end is indicated by a stop codon, either TAA or TGA. Since three bases code for each amino acid, there are three possible reading frames in each direction along the same sequence. The sequence can run in either of two directions, so that makes six possible reading frames

Table 9.1 Nucleotide triplet codes for the amino acids

1st base	*2nd base*	*3rd base* A	*C*	*G*	*T*
A	*A*	Lysine	Asparagine	Lysine	Asparagine
	C	Threonine	Threonine	Threonine	Threonine
	G	Arginine	Serine	Arginine	Serine
	T	Isoleucine	Isoleucine	Methionine	Isoleucine
C	*A*	Glutamine	Histidine	Glutamine	Histidine
	C	Proline	Proline	Proline	Proline
	G	Arginine	Arginine	Arginine	Arginine
	T	Leucine	Leucine	Leucine	Leucine
G	*A*	Glutamic acid	Aspartic acid	Glutamic acid	Aspartic acid
	C	Alanine	Alanine	Alanine	Alanine
	G	Glycine	Glycine	Glycine	Glycine
	T	Valine	Valine	Valine	Valine
T	*A*	STOP	Tyrosine	STOP	Tyrosine
	C	Serine	Serine	Serine	Serine
	G	STOP	Cysteine	Tryptophan	Cysteine
	T	Leucine	Phenylalanine	Leucine	Phenylalanine

for any piece of the genome. In some cases, the codes for different genes actually overlap one another.

The genetic code is highly redundant. Some amino acids have up to six different codes. Serine, for instance, results from each of the triples AGC, AGT, TCA, TCC, TCG and TCT. This redundancy occurs because there are just 20 amino acids, but 64 different possible triplets of bases in DNA. If you look at the table, you will notice that in most cases where an amino acid has several codes, it is repeated across an entire row. What this means is that the third base in the sequence has no effect on the product. The amino acid is determined by the first 2 bases. This property has led to speculation that, at some early stage of evolution, the genetic code consisted of couplets and that the third base was added at some later stage.

The mechanics by which biological species change and new species appear are governed by processes involving *genes*. Genes determine the character and identity of a species. During (sexual) reproduction, offspring acquire genetic information from both parents in equal proportions. This genetic information, which determines the biologic 'design' of the individual, is contained in a set of chromosome pairs (one from each parent). These chromosomes contain DNA sequences that govern the formation of proteins and control many aspects of growth. A gene is a sequence of nucleotides that have a particular function. The site where a gene occurs on the chromosome is called its *locus*. Different forms of a gene (e.g., blue versus brown eyes) are called *alleles*. Since both parents contribute genetic material, two alleles are found at each locus. Together they determine an individual's *genotype*. Species with two alleles at each locus are termed *diploid*; *haploid* species have only one set of chromosomes and therefore only a single allele at each locus.

A surprising feature of the genome is that most of it seems to be junk. There are over 3 billion bases in human DNA, but only around 30 000 genes. Less than 20 per cent of human DNA actually codes for genes. Not only is there a lot of junk DNA, but the genes themselves are fragmented. Often, a single gene is broken into many short coding segments, called *exons*. These are separated by junk sequences, called *introns*. There is increasing evidence that so-called junk DNA may not be rubbish at all. For instance, a study by Andrew Feinberg and his colleagues at Johns Hopkins University found that at least 4 per cent of this junk DNA is identical in mice and men.[5] Feinberg suggested that the shared DNA that lies close to genes probably plays a role in gene regulation. This makes sense when you realise that DNA molecules are huge. Not only are they helical, but on a larger scale they bend and twist into highly convoluted shapes. For genes to be read and copied, the structure must unravel and junk DNA may help to make this happen. Another factor is the need for adaptive flexibility. Junk DNA provides space in which genes can be copied, cut and spliced without disrupting existing parts of the code.

Cracking the code

Biotechnology is a science that has grown up alongside computers. A tradition of collating gene sequences and other crucial data in databases became entrenched from the beginning. This activity has created a whole new science called *bioinformatics*, which deals with the theory and practice of information methods in molecular biology. The ability to compare each new sequence with whole libraries of related sequences has added an entirely new dimension to molecular research. It has also led to entirely new kinds of study in which sequence libraries are searched, compared and interpreted. The similarities and differences between families of sequences provide clues to the structure, function and evolution of countless genes and proteins.

A key aspect of biotechnology is the ability to splice segments of DNA into an existing strand of DNA. In effect what happens is that you take the nuclear DNA from a bacterium and insert the gene that you want. If the inserted gene codes for a particular substance, then the genetically altered bacterium will start producing that substance. Gene splicing techniques have already been used to coax bacteria into manufacturing several key substances. A notable success is human interferon. This substance occurs naturally in humans and is important in maintaining resistance to infections. There are many cases where boosting levels of interferon would be of great medical benefit. Unfortunately, interferon cannot be manufactured by traditional means and the small amounts that could be extracted from natural sources made it prohibitively expensive. Gene splicing finally made it possible to manufacture interferon in relatively large quantities, making it a viable treatment at last.

As genomic data started to accumulate, researchers quickly realised that there were many advantages to be gained by comparing DNA sequences with one another. For instance, if the function of genes is known in a DNA sequence from one organism, then by comparing equivalent sequences from related organisms, you could make a reasonable guess as to whether or not the genes play a similar role. Similarities between different protein sequences provide hints about their properties, and often about their function. The so-called central dogma of bioinformatics is that similarities between different sequences of genes and proteins reflect similarities in their structure and function.

The most basic comparison between sequences is to align them. That is, you write them side by side so that matching bases (in DNA or RNA)

(a)

	L	A	A	R	L	A	W	T	Y	L	A	A	R	W	G	K	V	G
L	•				•					•								
A		•	•			•					•	•						
A		•	•			•					•	•						
R				•									•					
P																		
A		•	•			•					•	•						
K																•		
Y									•									
L	•				•					•								
A		•	•			•					•	•						
A		•	•			•					•	•						
R				•									•					
C																		
A		•	•			•					•	•						
G															•			•
N																		
V																	•	
G															•			•

(b)

	L	A	A	R	L	A	W	T	Y	L	A	A	R	W	G	K	V	G
L	**44**	37	31	29	29	22	22	22	21	21	14	8	6	6	4	3	2	0
A	38	**34**	29	26	25	21	20	20	19	18	14	9	6	6	4	3	2	0
A	33	30	**26**	23	22	19	18	18	17	16	13	9	6	6	4	3	2	0
R	31	27	23	**23**	21	17	17	17	16	15	11	7	7	6	4	3	2	0
P	31	27	23	22	**21**	17	17	17	16	15	11	7	6	6	4	3	2	0
A	26	24	21	19	18	**16**	15	15	14	13	11	8	6	6	4	3	2	0
K	25	22	19	18	17	14	14	**14**	13	12	9	6	5	5	3	4	2	0
Y	24	21	18	17	16	13	13	13	**14**	12	9	6	5	5	3	3	2	0
L	23	19	16	15	16	12	12	12	12	**13**	9	6	5	5	3	3	2	0
A	17	16	14	12	12	11	10	10	10	10	**9**	7	5	5	3	3	2	0
A	12	12	11	9	9	9	8	8	8	8	8	**7**	5	5	3	3	2	0
R	10	9	8	9	8	7	7	7	7	7	6	5	**6**	5	3	3	2	0
C	10	9	8	8	8	7	7	7	7	7	6	5	5	5	3	3	2	0
A	5	6	6	5	5	6	5	5	5	5	6	6	5	**5**	3	3	2	0
G	3	3	3	3	3	3	3	3	3	3	3	3	3	3	**3**	2	1	1
N	3	3	3	3	3	3	3	3	3	3	3	3	3	3	2	**2**	1	0
V	2	2	2	2	2	2	2	2	2	2	2	2	2	2	1	1	**2**	0
G	0	0	0	0	0	0	0	0	0	0	0	0	0	0	1	0	0	**1**

Figure 9.1 Alignments for segments of two proteins
*(*a) A simple dot matrix diagram. (b) Scores for the Needleman–Wunsch method of alignment. Entries corresponding to the best alignment are shown in boldface.

or amino acids (in proteins) in different sequences are placed next to one another. The earliest and simplest method of comparing two sequences was the *dot-matrix diagram*. This method treats two DNA sequences as headings for the rows and columns of a table. In the body of the table one enters a zero (shown as a blank in Figure 9.1a) if the headers for the row and column defining a cell are different and a one (shown as an asterisk in Figure 9.1a) if they are the same. Any diagonal row of asterisks indicates that those two sections of the sequences are identical. Of course, in reality there are usually differences. For instance, one DNA sequence may be missing strings of bases that the other sequence has, or it may have different bases at certain locations. So finding the best alignment can be very tricky. In 1970, two scientists, Needleman and Wunsch, used the dot-matrix idea to develop a way to find the best alignment of two related sequences. Their method adapts the dot-matrix diagram by scoring the number of matching positions (see Figure 9.1b). Simple dot-matrix methods are no longer as widely used these days, partly because modern researchers are likely to want to compare many different sequences at the same time. Researchers are more likely to use data

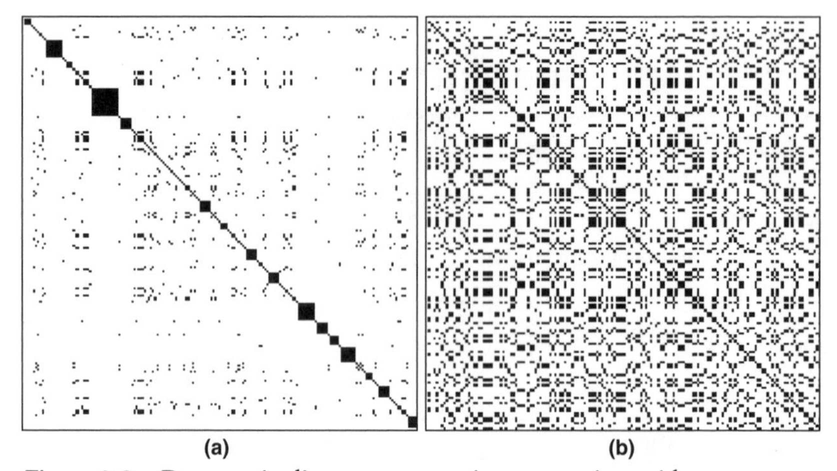

Figure 9.2 Dot matrix diagrams comparing two amino acids
(a) showing matches using the raw data (i.e., the amino acid sequences); (b) matches for the same sequences, but after replacing the amino acids with codes for their chemical properties.

mining methods such as those described in earlier chapters. Nevertheless, these more advanced methods still employ the ideas of comparison and similarity described here.

Unfortunately, the above methods become unworkable when many sequences are involved. So other methods are used. For instance, one approach is the idea of a *consensus sequence*. That is, you find a sequence that best represents the typical composition of a set of sequences. That done, you can then align new sequences against the consensus pattern.

There are obvious disadvantages to such an approach. For instance, each sequence may have some elements missing and others added. In a coding region of DNA, the obvious consensus pattern to use is the amino acid sequence that results from the code. In proteins, the properties of the amino acids determine how the protein folds, so a lot of research has focused on finding *homologues*, patterns that resemble each other in function. For instance, it may not matter which amino acid occurs at a certain point in the protein polypeptide chain, so long as it is (say) charged, or polar, or hydrophobic (Figure 9.2). This means that different sequences can actually fold into the same shape or serve identical functions.

A related feature is that proteins appear to be modular in structure. They are composed of many small building blocks that occur frequently. These short sequences of amino acids, called *motifs*, are typically sequences that fold in characteristic ways, such as loops, helices or sheets. They therefore influence the structure and function of proteins. Often the amino acids that fill in the gaps between motifs have no effect at all. The Prosite database lists thousands of protein sequences in terms of their motif patterns.[6] For example, the pattern [LIVFM]-x(2)-Y-D-S-[YA] describes a sequence of amino acids which starts with any one of the acids L, I, V, F, or M. The symbol x (2) means that there follows any two amino acids. So the sequences LACYDSY, VPQYDSA, and FWYYDSY all conform to this pattern. Motif searches can simplify alignment enormously because they reduce the number of combinations to be searched when trying to match different sequences.

Today, there is a very large range of computer-based tools available to help scientists interpret genes and proteins. Most of the main data repositories provide data processing services. These include tools for searching databases, for finding similar sequences, for aligning sequences, for interpreting their structure, for identifying their function and for finding taxonomic relationships among sets of organisms. Given the enormous collections of data now available, some of the simple methods described earlier have largely given way to machine learning, high performance computing and other advanced methods.

Recent developments in automation blur the line between chemistry and computing. A *micro-array* is a surface that is laid out as a biochemical matrix. Also called *gene chips*, or *DNA chips*, they consist of thousands of cells set out on a surface.[7] Each cell in the array contains a distinct oligonucleotide (gene fragment). They are used to determine which genes turn on or off under certain conditions. Micro-arrays can often contain literally thousands of cells—they are such complex structures that the arrays are usually printed in a grid, in much the same way as printers output images as a collection of pixels.

The grid in a micro-array can contain tens of thousands of cells, each containing a gene arrayed at a specific position.[8] The micro-array allows researchers to carry out many experiments all at once. The results can be automatically read using scanning methods and then stored in computer databases. In effect, the micro-array becomes a kind of input device that registers genetic responses.

From gene pool to gene bank

As the number of genomic sequences increased, people realised that some sort of reference library needed to be established so that scientists could compare new sequences against known ones. The importance of such a database, and the need for information processing to interpret the data, meant that special organisations were set up to collate the data and to develop the methods.

One of the successes of biotechnology has been the compilation of comprehensive data banks. The largest international databases, for instance, include contributions from thousands of different sources. The International Nucleotide Sequence Database Collaboration (INSDC) comprises the world's three main biotechnology databases: GenBank in the USA, the European Molecular Biology Laboratory (EMBL) and the DNA Data Bank of Japan (DDBJ).

GenBank is a public database of all DNA, RNA, and protein sequences.[9] It started in 1982 with 606 entries. By 1988, administrators realised that a specialised national centre was needed to manage the future growth of genomic information. The US Congress passed legislation creating the National Center for Biotechnology Information (NCBI) on 4 November 1988. The National Library of Medicine (part of the National Institutes of Health) was already creating and maintaining biomedical databases, and was therefore chosen to build and run the NCBI. Besides carrying out research in genomic modelling and analysis, the centre has also developed automated information systems to support molecular biology and biotechnology.

The DNA Data Bank of Japan (DDBJ) opened in 1986 with endorsements by the Ministries of Education, Science, and Sport and Culture.[10] At the end of its first year, it held just 66 entries, a slow start, but it accelerated during the 1990s, reaching nearly 10 million by 2000.

Unlike the other two databases, which are each based in a single country, the European Molecular Biology Laboratory (EMBL) is itself an umbrella for cooperation between sixteen countries, mostly in western Europe, but also including Israel.[11] It was established in 1974 and includes five facilities: the main laboratory in Heidelberg (Germany), outstations in Hamburg (Germany), Grenoble (France) and Hinxton (the UK) and an external Research Program in Monterotondo (Italy).

Today, the three international databases are almost identical in content. To ensure that authors have access to the most up to date information, the databases swap updates with each other on a daily basis. By the end of 2000, they each held records for about 9 million sequences (over 10 billion bases), and include sequences from more than 50 000 species. The terrific growth in biotechnology is reflected by the increasing size of the databases, which have been growing almost exponentially. The DDBJ, for instance, reached 1 million entries at the end of 1996. By the end of 1997 it had increased to nearly 2 million and was nearly at 3 million by the end of 1998. It passed 4 million at the end of 1999.

These bioinformatics projects not only promote collaboration on a truly vast scale, but also enrich research. One advantage is that they make possible entirely new kinds of research, especially data mining. The serendipity effect is important because combining different kinds of data inevitably leads to new and unexpected discoveries. For instance, it is now standard practice to compare newly sequenced genes against known sequences in the major international databases. There have already been several studies attempting to identify families of genes by doing structural comparisons of entire databases.[12]

At first, the databases were simply services for molecular biologists. However, as their usefulness grew, and as everyone took to using them, contributing to them became compulsory. Top-ranking journals such as *Science* and *Nature* will not accept a letter or article for final submission unless the data has been submitted to an international database. Many funding bodies now insist on the same thing.

By the end of the year 2002, molecular biologists had put together complete DNA sequences for several species. These include those listed in Table 9.2, plus 30 or so microbes.

Among the first species to be sequenced were organisms that had figured prominently in scientific experiments, including the bacterium *Escherichia*

Table 9.2 Sizes of some genomes sequenced by end of the year 2002

Name	*Millions of bases*	*Number of genes (approx.)*
Mouse (Mus musculus)	3000	30 000
Human (*Homo sapiens*)	3000	30 000
Fruit fly (*Drosophila melanogaster*)	135.6	13 061
Thale cress (*Arabidopsis thaliana*)	100	25 000
Nematode roundworm (*Caenorhabditis elegans)*	100	13 000
Brewer's yeast (*Accharomyces cerevisiae*)	12.1	6034
Escherichia coli	4.67	3237
Bacteria (H. influenzae)	1.8	1740

Source: Human Genome Project http://www.ornl.gov/hgmis/faq/

coli and the fruit fly (*Drosophila melanogaster*). In 2000, the experimental organism Thale cress (*Arabidopsis thaliana*) became the first plant to be sequenced completely. Early in 2001, rice became the first food crop to be completely sequenced. Given the importance of plants as food sources, complete sequences of more plants are likely to follow quickly. The gene sequence of the mouse was completed in 2002. Not only is the mouse a much-used experimental animal, it is also the first mammal other than humans to be sequenced. As Table 9.2 shows, its genome is similar in size to humans. Being a mammal, the mouse genome should have revealing similarities to the human genome.

It is essential to get complete sequences for a wide range of species, for several reasons. First, scientists want to have sequences for economically important species. Second, they want to understand how the human genome works. They can get vital information from the similarities and differences between sequences for different species. For instance, similarities between the equivalent gene in different organisms would help to identify the features that are essential to its function. The same idea applies to similarities between *genetic pathways* (sets of interacting genes). Such research would help to identify (and ultimately even design) new and useful genes, such as those conveying disease resistance in crops.

The Human Genome Project

Mention 'big science' and you immediately think of projects like the building of the first atomic bomb, particle accelerators, giant telescope arrays or putting man on the moon. Until 1990, there was really nothing comparable in the life sciences.

The Human Genome Project (HGP) aimed to produce a complete listing of the entire human genome, including all of the 3.2 billion base pairs in its DNA sequence, and a complete description of all human genes.[13] The project began formally in 1990 and aimed to complete mapping the human genome within 15 years. Until 1998, progress was slow and it looked as though the target date might not be met. However, accelerating the rate at which DNA sequences can be read allowed the project to obtain a complete draft of the human genome by June 2000.

Right from the start, it was clear that speed would be the essential factor. The human genome is so big that even with dozens of labs participating, it was going to take decades. One advance was automation. At first, the data for even short DNA sequences was hard to obtain. However, people soon designed and built machines to carry out the entire process automatically.

Another issue was the basic sequencing strategy. Craig Venter proposed a radical shotgun approach, in which an organism's DNA is blasted into thousands of fragments. These fragments are then sequenced individually and the results are reassembled.[14] Venter had already used the shotgun method successfully at The Institute for Genome Research (TIGR) to sequence the influenza genome. However, HGP scientists were sceptical that the method would work on anything as large as the human genome. They also had much time and effort invested in a slower, but systematic tiling approach.

In 1998, Venter announced that the entire human genome would be sequenced within three years. He founded a new company, Celera Genomics Corporation, and began work on the human genome independently. From then on, the rate of sequencing accelerated rapidly. A draft of the human genome was completed several years ahead of schedule.

Making sense of it all

Sophisticated as they were, the early applications of cutting and splicing genes were relatively primitive. Essentially, they relied on a simple 'cause and effect' principle. That is, you find a single gene that makes a certain product and you cut and paste it into a microbe. This microbe then becomes a little factory, churning out that product in great quantities. Now don't get the wrong idea. This approach can yield some powerful applications, for example human interferon (see page 134).

Powerful as they most certainly are, applications of simple splicing are only the beginning of the story. The classic features that genetics deals with, such as eye colour, are each controlled by a single gene. But to understand the formation of the eye itself, we need to consider how the activities of many genes are coordinated.

One of the greatest mysteries in biology is how the genetic code influences the development and growth of organisms. How do genes turn growth on and off at the correct time? An important discovery was the finding that some genes influence others. Controlling the operation of other genes (that is whether they are on or off), seems to be the main role of an entire class, called hox genes.

Francois Jacob and Jacques Monod were awarded the Nobel Prize for showing that genes can be organised into functional modules or complexes, which they called *operons*.[15] In *E. coli*, the metabolism of lactose depends on a complex of genes (an operon) that produce the necessary enzymes. The operon is activated by a regulatory gene which turns the genes on and off. Subsequent research suggests that many genes function as components of modules and are turned on and off by regulatory genes. For instance, in humans and other animals, the development of the eye is regulated by a single 'eyeless gene', which prevents the suite of genes involved from developing eyes except when needed. Experiments have shown that turning this gene off leads to the development of eyes on many parts of an insect's body.[16]

Such results imply that regulatory genes form a large network. To understand the complexity of growth, therefore, we need to understand how regulatory networks are organised and how they behave. In the late

1970s, Stuart Kauffman modelled genetic control over development in terms of structures called *Random Boolean Networks* (RBN).[17] In this model, each gene is like a switch and is either on or off. Besides having some primary role (such as eye colour), genes can also affect one another. So when one gene turns on, it might also turn off some other gene.

The RBN is an abstract model that incorporates only the barest minimum of details about genes. It contains no information about DNA coding or protein production. Nevertheless, it does capture some important properties of genetic control. Kauffmann's research turned up some interesting results. For example, he found that the models display cyclic behaviour and the cycling period increases in relation to the size of the network. This pattern may explain a similar link between genome size and the time separating one generation from the next. The model also points to properties that were previously unsuspected. For instance, he demonstrated that the ability of such systems to evolve is linked to the formation of *frozen components* (groups of nodes that are permanently on or off).

Research into genetic regulation is in its infancy. With details of entire genomes now available, the challenge is to flesh out the details of exactly how the genome controls growth and devlopment. Ultimately, this work holds the potential to predict exactly how any change in the genetic makeup of a plant or animal will express itself in the adult. This raises all kinds of possibilities, such as designing organisms.

A genetic cornucopia

The biotechnology revolution relies on information. Whether referring to type culture collections or comparing gene sequences, access to comprehensive, up-to-date biological information is essential in almost every aspect of biotechnology. The impact of biological databases has had perhaps as great an impact as (say) cloning and other breakthroughs in laboratory techniques.

Biotechnology has been hailed as the major growth industry of the twenty-first century. With this in mind, bioinformatics has the potential to become a veritable multi-billion dollar cornucopia.

Table 9.3 Some biotechnology resources available on the Internet

American Type Culture Collection (ATCC)	http://www.atcc.org/
DNA Data Bank of Japan	http://www.nig.ac.jp/index-e.html
Enzyme Data Bank	http://kr.expasy.org
European Molecular Biology Laboratory	http://www.embl-heidelberg.de/
The Genome Database	http://www.gdb.org/
GenBank	http://www.ncbi.nlm.nih.gov/
EXpasy	http://kr.expasy.org/
Protein Data Bank (PDB)	http://www.rcsb.org/pdb/
Restriction Enzyme Database (NEB)	http://rebase.neb.com/rebase/rebase.html
WFCC World Data Center on Microorganisms	http://wdcm.nig.ac.jp

Note: Online addresses change very quickly. These addresses were correct at the time of writing.

The number of biotechnology resources on the Web is increasing rapidly. Some of the major resources, plus their respective network addresses, are listed in Table 9.3. By far the biggest resources are databases, especially compilations of genomic sequences, enzymes, type cultures and taxonomic records, but there are also many large repositories of useful software. Most of the prominent sites provide a wide range of other information as well, including bibliographies, electronic newsgroups and educational material. There are also many pioneering services that explore new ways of using the network.

Bioinformatics is the science of applying computers to biological information. However, the term has come to be most closely associated with the role of computers in biotechnology. Most obvious is the use of computers to store data, especially gene and protein sequences. However, computers are also essential tools for interpreting this data.

The technologies described above make an extremely powerful combination. Like other fundamental technologies, such as the wheel or the production line, they can be turned to many different applications. They raise the prospects of manufacturing a huge variety of exotic substances, using the same generic technique to synthesise each one. In principle it works as follows.

Assuming you know the genetic sequence that codes for the substance you want, the ability to synthesise strips of DNA to order allows you to produce a strip of DNA from the sequence data. You then splice that strip into a bacterium and culture it. The bacterium then manufactures the substance for you.

Take note of that big 'if' in the above description—you can manufacture a substance *if* you know its DNA sequence. This need highlights the issue that, as a commodity, genetic information has potentially enormous value. Companies are now racing to identify the genetic sequences for all manner of substances that may have commercial value; a race not restricted to human DNA, of course. There are any number of currently rare or expensive substances produced by other organisms that could be manufactured biochemically.

The possibilities do not end there. Far from it. For a start, not only can you manufacture natural substances, you can also manufacture entirely new artificial ones. To do this successfully depends on developing an understanding of the relationship between structure and function. The way a chemical compound acts is governed by its physical properties. For large molecules, these properties are largely related to its three-dimensional structure, such as the arrangement of bonding sites, areas of charge, and so on. In some cases biochemists know enough about these relationships to pinpoint the exact effect different features have. It is then possible to see how to alter an existing substance to achieve a desired effect. As the knowledge of structure and function increases, it will be possible to design completely artificial substances.

The stuff of dreams or an ethical nightmare?

One of the biggest problems that we face in trying to deal with complexity in the world is that you cannot isolate one problem from another. Inevitably, people's concerns about different issues are going to come into conflict with one another, which is when the need for trade-offs arises.

To take an example, opposition to genetically modified foods stems from concerns about their unknown and possibly devastating side effects. One modification that has been much discussed is genetically engineering crops to include a built-in resistance to disease and insect damage.

Genetic alteration poses some hazard. The modified crops could have unexpected lethal effects on humans, or they could escape and become serious weeds. It has several potential benefits, however. If plants repel insects, then there is no need for pesticides. The risk of environmental side-effects, such as poisoning wildlife, is removed.

Which is the greater evil? Which is the greater good?

In his novel *Brave New World*, Aldous Huxley described a hedonistic society dominated by technology. It is a world in which current social values have been turned upside down. For example, foetuses are grown in test-tubes and genetically manipulated to produce adults with predetermined intelligence and social standing. The technology to do this may not now be far away, but Huxley does not offer a solution. His hero, unable to cope with the system, commits suicide. Huxley simply asks the question, and delivers the warning.

Would-be parents can now select from a smorgasbord of sperm choices, including sports heroes and Nobel Prize winners. However, as our knowledge of the structure and function of the human genome grows, genetic engineering raises the spectre of more bizarre possibilities. For a start, there is already much talk about genetic screening of embryos. That is, foetuses could be checked to ensure that they have no abnormalities. The implication is that parents might seek to terminate pregnancies if they know that their baby will be less than perfect. In time, it might be possible to circumvent this ethical problem entirely. Human eggs or embryos might be harvested, much as they are today, and then undergo genetic enhancement.

There are many computer games in which you choose the hero or heroine who will represent you in the game. Often, you can not only change the clothes that your player will wear, but also mix and match their appearance by choosing height and build, as well as hair and skin colour. It is conceivable that, in the not too distant future, having a baby will be a similar process.

Consider this scenario. The setting is the future, many years in the future for sure, but perhaps within our lifetimes. A young couple wants to start a family. Now at this point, in the good old days, they would simply go to bed and get on with it. However, in this brave new millennium they go to their gene therapist instead. From samples of their DNA, the therapist puts together a sketch of the range of prospective children they could have. Possible defects are noted, and marked for correction. Of course, they would like the child to look like them, but they also want him or her to have the best start in life. So they ask the therapist to alter various features by snipping in key genes from the spectrum available. The new genes do not actually have to come from another third person at all. They can be composed on the spot from a library of sequences.

chapter 10

THE INTERNET TURNS GREEN

The massive development of biodiversity-related information systems on the Internet has created much that appears exciting but chaotic, a diversity to match biodiversity itself.

Professor Frank Bisby, Chairperson, Species 2000[1]

Towards the end of 1983, an Australian Air Force F-111 bomber plane took off on one of the strangest missions in the country's military history. Flying low over the wilderness in Tasmania's southwest, it took a series of photographs of the Gordon and Franklin Rivers, then headed back to base. The plane had just completed a spying mission—on its own territory! The problem was that the Commonwealth government of Australia was in the midst of a dispute with the state government of Tasmania. The state government wanted to build a hydroelectric dam which would flood the Franklin River, an important conservation area. The Commonwealth government had nominated the region for listing as a World Heritage area and wanted to stop the dam from being built. At the height of this dispute, the state government refused to share important information with its national counterpart. For weeks, the state police had been preventing anyone from entering the area. So the Commonwealth's only way of getting the information it needed was to take photos from the air. Because of the gung-ho approach it represented, this internal spying mission became infamous as the 'Biggles incident', after W.E. Johns' fictional pilot hero.

Extreme measures, such as spying on your own country, become necessary when a government does not have the information it needs at its fingertips. Lack of adequate information still hampers environmental planning and management.

Seeing the wood for the trees

One day, not so long ago, I was presenting a public lecture about the efforts to compile biodiversity databases around the world. Afterwards a member of the audience stood up and complained that we computer types were fiddling while Rome burned. The money, he believed, should be spent on action, not numbers. His reaction was typical of many frustrated environmentalists, who cannot see the point of environmental databases. Another common reaction is surprise. What can computers possibly do to save the wilderness and wild animals?

The answer, as it turns out, is that computers have a lot to do with conservation. Governments have learnt from bitter experience that good information is crucial to good decision-making. As the worldwide environmental movement got into its swing during the 1970s and 1980s, governments found themselves lurching from crisis to crisis. There was no basis for decision-making—whether it was pulp mills in the wilderness or building roads through rainforests, every decision had to be made *ad hoc*. Every solution was, at best, a band-aid.

Another frustrating aspect of the period was that decision-makers had no way of knowing what data to believe. The problem was most acute during logging disputes. Conservationists would say one thing; loggers would say another. In Australia, this frustration led to the 'Biggles' spying incident described on page 147. Fed up with the confusion many countries decided to set up their own environmental information agencies to help put decision-making on a more systematic and informed basis. As a result, Australia now has environmental information offices in every state—as well as the Commonwealth. Many other countries have also set up environmental information systems in response to similar experiences.

The scale of the problem

One of the great challenges facing us at the start of the new millennium is how to conserve and manage the world's living and natural resources. As the human population grows, so does the pressure on resources. We have now reached a point where almost no place on earth is untouched by human activity. It is by no means clear that the world's resources can sustain such a large mass of people indefinitely. Slowly we are learning to use our resources more carefully.

Looking at the worldwide picture, the challenge of documenting the environment is mind-boggling. The planet's surface area exceeds 509 000 000 square kilometres. Simply monitoring such vast tracts is a huge task in itself. Then there is the problem of documenting the world's plant and animal species. Over the past 200 years, taxonomists have described some 1.5 million species of plants and animals. But this is only the beginning. The total number of species is not yet known, but is estimated to be somewhere between 10 and 100 million. At the current pace of description, it could easily take over a century of taxonomic research simply to document them all. Modern technology can help with this task but at the same time it generates huge quantities of data that must somehow be stored, collated and interpreted.

Given the size and urgency of the problem, piecemeal solutions simply will not do. Humanity as a whole needs to plan and act systematically, but governments need sound, comprehensive information on which to base their plans. The problem is so huge that nothing less than the coordinated efforts of every environment and resource agency in every country will be adequate. The Internet is certain to play a vital role in achieving this goal.

The changing nature of conservation

For most of the twentieth century, 'conservation' implied parks and reserves. The rapidly growing scale of environmental alteration, however, and increasing public awareness of environmental issues, has highlighted a need for off-reserve conservation and broad-scale landscape management. A few of the resulting issues include: environmental impact assessment, state of the environment reporting, environmental monitoring, conservation of rare and endangered species, natural heritage planning, species relocation programs, land use planning and environmental degradation. Local decisions and priorities need to be set in a wider, and ultimately global, context. For instance to decide whether or not to log a patch of forest, planners need to know how much forest there is, what species might be put at risk, what the global costs and benefits are, and so on. Conversely, every local area contributes valuable data and experience that can feed into global planning and be applied to other areas.

The informatics-based approach that has emerged treats environmental management as a host of conservation activities that reinforce each other, but are constrained by the global picture. Good communication is crucial. In terms of data sharing, this means everyone contributing data, which is then made widely available. Setting matters in context means having access to relevant and reliable information. By the 1990s, governments were very active in setting up regional, national and international environmental information systems.

From maps to virtual geography

Geographic information systems (GIS) are computer programs for acquiring, storing, interpreting and displaying spatially organised information. GIS had their origins in many different disciplines, including electronic cartography, geological surveys, environmental management and urban planning. It has now become an essential tool in all of these professions, as well as many others.

At the time of writing, a great transition is taking place in the way geographic information is handled. Instead of being isolated in stand-alone machines, a new environment is being created in which geographic information is stored and accessed over the Internet. This transition has implications that go far beyond just a change in format, one of which is the greater accessibility that the Internet offers. However even more significant is the potential to combine information from many different sources in ways that were never previously possible. To work in the new environment, managers, developers and users all need to learn the basic technology involved.

Geographic data consist of *layers*. A layer is a set of geographically indexed data with a common theme or type. Examples might include coastlines, roads, topography, towns and public lands. To form a map, the GIS user selects a base map (usually a set of crucial layers, such as coastlines or roads) and overlays selected layers.

Strictly speaking, a geographic information system consists of three main components: *data*, *processing* and *user interface*. In a 'traditional' GIS, all three elements sit inside a single computer on the user's desk. In an online system, these elements may all reside in different computers in different places.[2] A 'traditional' GIS package may be Internet aware—able to download and use data sets from remote sites. Most commercial GIS packages now include Internet features. One advantage of this is that specialised layers or queries can be accessed and used as they are needed. Alternatively, a GIS engine may sit next to a Web server and deliver geographic information services to remote users, via a standard Web browser. This makes simple GIS services widely available in a cost-effective manner, as the users no longer require expensive hardware and software. Many GIS engines now support delivery of spatial data via a Web server.

The technology for handling GIS online is evolving rapidly. One of the greatest needs at present is to develop suitable standards for sharing geographic information over the Internet. Organisations such as the Open GIS Consortium are developing specifications for the creation of open networks.[3] The rise of XML (see Chapter 6) has spawned many relevant languages and standards, such as the Geographic Markup Language, GML.[4]

The logical endpoint of putting geographic information online is to create a comprehensive, global GIS. If information from different sources can be combined seamlessly into a single resource, then in principle there is nothing to prevent the development of such a system. Many of the relevant services already exist, most providing raw material that could become components of a global GIS.

The first online mapping service was Xerox PARC's *Map Viewer*, which opened in 1993.[5] Most online GIS systems are based on the *Digital Chart of the World*, a collation of coastline and other geographic data that was compiled by the military. Using any one of dozens of online services it is possible to draw maps to a resolution of 1 kilometre or better, for anywhere in the world. Other sites now provide detailed coverages or queries for particular areas or particular themes, such as physical, biological, economic and political data.

The challenge to produce a global GIS online is not a new idea. There are plenty of precedents to work from as integrated online services, many of them global, already exist. They prove that a comprehensive global GIS is a practical possibility.

Probably the best examples to consider are services that deal with tourist information. In almost every case, the service provides systematic indexes that link to large networks of online sources of geographic information. Many of these geographic networks, such as geographic indexes of hotels and other accommodation, are highly specific and are sometimes subsidised by a particular industry group. However, some networks have taken a broad brush approach from the start. Prominent examples include the *Virtual Tourist*, and the Lonely Planet travel guides.

Perhaps there is no need to build a global GIS from scratch if such systems already exist. Some of the services mentioned above are very impressive in the range and depth of information they supply, but most are responding to particular commercial needs and opportunities.

Some concerns are already being covered by international cooperation between governments. For instance, the Global Biodiversity Information Facility (GBIF) is the result of international agreements on biodiversity conservation.[6] Such networks are still responding to a perceived need, but this time environmental, rather than commercial.

So why is a global GIS needed? One answer is that in an era of increasing globalisation, people need to be able to access and combine many different kinds of detailed information from anywhere on Earth. Also, another aspect of globalisation is that almost every activity impinges on everything else. For example, the proponents of a commercial venture need to know about a possible site's environment, social frameworks and politics so they can minimise possible impacts and repercussions. Likewise, in conservation, managers and planners need to be aware of the potential commercial, political and other consequences of banning development in particular regions.

The role of the Internet

The rise of the Internet, and especially the World Wide Web, revolutionised the dissemination of biodiversity information during the 1990s. Once an organisation had published information online, anyone, anywhere could access it. Users could access relevant information more readily than before. They could also access it faster and in greater quantities. This raised the possibility of organisations providing general access to information resources that normally require specialised software or hardware. For example, remote users can query databases by filling in an online form. Geographic information systems used to require specialised, often expensive, equipment. Now a standard Web browser is all the user needs.

To realise the potential for a global geographic information system international coordination is essential. Organisations and nations need to agree on protocols and standards for data recording, quality assurance, custodianship, copyright, legal liability and indexing.[7] International agreements, such as the Convention on Biological Diversity, helped to set this process in motion.[8]

How do we organise information on a large scale? One approach is to start at the source and organise publishing sites into an *information network*. In this context, an information network is a set of sites on the Internet that coordinate their activities. The sites operate under a common framework, especially in the indexing of the information that they supply. The advantage of information networks is that they can address directly issues that are crucial in building a reliable information system, including: standardisation, quality control, indexing and stability.

Information networks for biodiversity

The Internet creates the potential to develop worldwide biodiversity information networks. As the World Wide Web spread, the 1990s saw a proliferation of cooperative projects to compile biodiversity information online.

One useful outcome of all this networking activity is that a lot of biodiversity information, such as taxonomic nomenclature and species

Table 10.1 Some biodiversity services and organisations on the Internet

Organisation	*Web address*
CIESIN	http://www.ciesin.org/
Convention on Biological Diversity	http://www.biodiv.org/
DIVERSITAS	http://www.icsu.org/DIVERSITAS/
Environment Australia	http://www.environment.gov.au/
European Environment Information and Observation Network (EIONET)	http://www.eionet.eu.int/
Global Biodiversity Information Facility (GBIF)	http://www.gbif.org/
International Legume Database & Information Service (ILDIS)	http://biodiversity.soton.ac.uk/LegumeWeb
International Organization for Plant Information (IOPI)	http://plantnet.rbgsyd.gov.au/iopi/iopihome.html
International Union of Forest Research Organizations (IUFRO)	http://iufro.boku.ac.at/
Species 2000	http://www.species2000.org/
Tree of Life	http://tolweb.org/tree/phylogeny.html
United Nations Environment Programme (UNEP)	http://www.unep.org/
USDA Plants Database	http://plants.usda.gov/
World Conservation Monitoring Centre (WCMC)	http://www.unep-wcmc.org/

Note: These addresses were correct at the time of writing.

checklists, is now online. A number of international projects have made this a priority. For instance, in 1992 the International Organization for Plant Information (IOPI) began developing a checklist of the world's plant species.[9] The Species 2000 project has similar objectives.[10] At the same time, the Biodiversity Information Network (BIN21) set up a network of sites that compiled papers and data on biodiversity on different continents. As we can see in Table 10.1 there are now many online information networks that focus on environment and resources.

Established institutions also have seen the need to place their data resources online. In particular, most major museums and herbaria have invested in programs to set up databases for their collections. For many of these institutions, this process has not been easy as funds to catalogue obscure plants and animals are not in great supply. It is an essential job, but not very glamorous.

It is also a huge undertaking. Some large museums house up to 10 million specimens, collections that have taken over a century to put together. If you hired a full-time technician to enter the data from specimen labels into a database, after a year's hard work, you might have a corpus of 50 000 entries. This sounds impressive until you realise that, at the same rate, it would take 200 years to complete the database! And what about the millions of specimens that might accumulate in the meantime? Given these realities, there is considerable pressure on systems designers to come up with efficient alternatives. For instance, as well as handing in specimens, field biologists now typically fill in an electronic form, so the database can be updated directly.

The greatest challenges in collating biodiversity information have been human—especially legal and political—issues, rather than technical problems. One outcome of the Convention on Biological Diversity was the concept of a Clearinghouse Mechanism.[11] In the short term, this scheme aimed to help countries develop their biodiversity information capacity. The longer-term goal was to enhance access to information through a system of clearinghouses which would gather, organise and distribute biodiversity information. The next challenge will be to merge these clearinghouses into a global network.

Towards a global information system

Until recently, the idea of a global warehouse of environmental information was unthinkable. It simply is not possible to collate all the available information in one place. However improvements in communications, and the rise of the Internet as a global communications medium, now make it feasible that such a system could be designed as a distributed network of information sources.

During the 1990s, concerted worldwide efforts began to create environmental information systems. CORINE, set up by the European Community, provided a common information base for the emerging European economy. This venture immediately highlighted some of the practical difficulties of combining data from many different sources. For example, each country had its own maps, which differed in both scale and accuracy. They also had different legal requirements about the provision and use of the data.

In 1993, the World Wide Web began to play a significant role in the attempts to develop environmental information. The Australian Environmental Information Resources Network attracted worldwide attention when it made continental scale information available on the Internet.[12] At the same time people began to experiment with information networks on the Web. A good example was the Tropical Database (BDT), a small research institute in Campinas, Brazil, which suddenly became a major international centre for biodiversity. In 1993, BDT organised a series of workshops about biodiversity information. Besides inviting participants to visit their institute, they made what turned out to be a crucial decision. They invited people to become virtual participants by submitting 'talks', comments and questions via the Internet. Immediately the event was swamped with contributions from all over the world. These workshops led to the formation of the Biodiversity Information Network (BIN21).

Another crucial development was the emergence of information networks. One of the real strengths of the Web is its ability to merge information from many different sources seamlessly. A single index, for instance, can point to items in hundreds of different locations all over the world. Scientists quickly realised that by coordinating their efforts they could create a far more accurate, comprehensive and up to date information system than they could as individuals.

In 1994, the OECD set up a Megascience Forum to promote large science projects of major international significance. The Human Genome Project was one such enterprise; another was the proposal for a Global Biodiversity Information Facility (GBIF). The aim of GBIF is to establish '. . . a common access system, Internet-based, for accessing the world's known species through some 180 global species databases.'[13]

From environmental data to information

The data that is now available is worthless unless people can use it effectively. Along with data warehouses, we also need information systems to interpret and apply the information. For instance, foresters, faced with the need to demonstrate the environmental impact of logging operations, developed simulation tools such as the visualisation program *SmartForest*. This program integrates simulation models with geographic information to create views of future landscapes under selected scenarios.

Despite all the successes of environmental informatics so far, there are still numerous problems to be solved, one of the most important of which is to mobilise all sources of information. To date, most of the activity has been confined to large government-sponsored organisations although the bulk of raw data is still produced by individuals in the course of research projects.

Here is one area where we still see a lot of waste. Much of this data is collected piecemeal for particular research projects. As soon as a project is complete and the results sent off for publication, potentially valuable data is left lying on someone's shelf. This means that the community is not getting full value for research funding. Besides contributing to individual studies, every piece of data could also help to fill in a larger environmental jigsaw.

To see the potential value of such activity we need to look no further than bioinformatics, which we looked at in Chapter 9. Bioinformatics has been so fruitful that many biotechnology companies are making the bulk of their data freely available. Research journals and funding agencies now make it compulsory for researchers to submit their data to appropriate public databases.

The principle of data recycling can and is being extended to many areas of study, especially environmental sciences. The main problem is that the crucial data has to be precisely identified—for example, a DNA sequence is an obvious unit of data. Another problem is that people measure the environment in many different ways for different purposes, so there has to be a set of consistent standards. Just setting consistent taxonomic references is still a nightmare for field workers.

Nevertheless, some kinds of biodiversity information are obvious. The most basic data are records of the type 'species X at point Y at time T'. So one important database would be a compilation of site records that combine species lists with basic site features such as elevation, climate, soils etc. When people started compiling environmental data from existing sources, however, many annoying problems came to light. The first was the poor quality of most data. For instance, when records were first compiled for rare and endangered species of trees in Australia, the resulting maps showed many sites out in the ocean. Now trees do not grow in the ocean; the maps were revealing errors in the data. Another annoying problem is the sampling bias: draw a map of species records for central Australia and you have a perfect map of the main roads across the continent. There are vast areas that no-one has ever surveyed.

Organisms and landscapes

Where would you find a spotted owl? Obviously, in the kinds of places where spotted owls are known to live. All biogeographers know that environmental factors limit the distributions of plants and animals. A number of scientists have successfully used climatic parameters as a means of predicting where particular species might be found[14]. To do this, you look at the places where a species is known to live and note the environmental characteristics of those locations. These might include temperature range, rainfall, soil types, and so on. Having done that, you then look for other sites that fit the same profile. These sites indicate the potential range of the species.

The above exercise is of more than academic interest. If you are trying to conserve an endangered animal, for instance, it helps if you can predict where you might find it. It also helps if you can identify areas that might be suitable for relocation programs. There are also commercial applications. If you are thinking of farming a new crop, then it would help to know what local or introduced insects might threaten your harvest.

These methods work well, up to a point. But there are limitations. It is difficult (and, with pest species, dangerous) to assume that knowing where a species is *not* found tells you what conditions it cannot tolerate.

Just because a plant is not found in an area does not mean that it could not grow there—think of the vast numbers of exotic plants that are cultivated in people's gardens around the world. Another complicating issue is that few environments these days are in a natural state of balance. Human actions have a big influence, especially when they alter the landscape and its environments. Finally, there are inter-species interactions. Plants and animals compete with each other for resources. In a landscape, this limits the range over which each species is found and is responsible for many abrupt transitions, such as borders between forest and grassland.

The future of conservation

In this chapter we have seen that conservation is changing. It has now been recognised that conservation activities need to be everywhere, not restricted to creating and maintaining reserves. Also, conservation activities need a rich supply of information about species, about environments and about conservation activities. All over the world, environmental planners and managers are tapping into the resources of the Internet to share information and to coordinate their activities.

Conservation is not simply a matter of preserving individual species. Interactions between species, their landscapes and environmental factors make ecosystems truly complex. To understand these ecosystems, and to predict the effects that various activities and management practices are likely to have, scientists and managers need to be able to perform virtual experiments in virtual worlds. In the next chapter, we will look at some of these virtual worlds.

chapter 11

VIRTUAL WORLDS

... In real life mistakes are likely to be irrevocable. Computer simulation, however, makes it economically practical to make mistakes on purpose. If you are astute, therefore, you can learn much more than they cost. Furthermore, if you are at all discreet, no-one but you need ever know you made a mistake.

John McLeod and John Osborn[1]

Fun and games

When the company Sony released a new games computer on to the market in 2001, there were scenes of mayhem in stores worldwide as stocks sold out within minutes. Such is the drawing power of games technology. From a pastime for programming buffs, computer games have become a multi-billion dollar *a year* industry.

Many computer games place the user in a virtual world. In games such as the *Sims*, *SimCity* and *SimEarth*, for instance, the user designs and builds entire communities, cities, even an entire planet. The games are based around a Geographic Information System (GIS) that allows the user to select regions and zoom in to the level of individual buildings. In other games, players can drive through virtual landscapes following road maps (e.g., *GranTurismo*) or negotiate their way through three-dimensional artificial worlds (e.g., *Tomb Raider*). The next logical step is to imitate the real world itself inside a computer.

Having combined GIS and virtual reality, it is just a short step (conceptually at least!) to do the whole thing online. Perhaps the best known example is *AlphaWorld*.[2] Participants can join a virtual community inhabited by *avatars*; graphical representations of both themselves and other users. Not only can users move around within this cyber world, they can even build themselves virtual homes. Other online projects of this kind place the avatars in alternative environments such as the surface of Mars. The Virtual Worlds Movement aims to use online virtual reality as a means of developing virtual communities and other, more mundane activities such as business meetings.[3]

If users of the Internet can explore alternate universes online, and interact with other explorers in the process, then why not explore virtual reproductions of the real thing? Many areas, such as city centres, are already mapped in detail. Tourist maps sometimes include pictures of famous buildings. Why not turn these pictures and maps into online virtual worlds? Tourists could get the feel of their destination before they even get on a plane, or explore remote areas such as the Rocky Mountains or the Andes in greater depth (and greater safety) than they ever could in real life.

More serious games

One day, in the midst of explaining the latest developments in information technology to visitors, I began to demonstrate our latest and greatest virtual reality software. Although this software was developed for a serious purpose: researching scenarios for environmental management, my audience was not impressed. They immediately dismissed it, saying 'We're not interested in computer games, what serious research are you doing?' This was not the first time that I've seen people confuse the cutting edge with the trivial. But it demonstrated a strange irony of computer science: the most advanced research and development is often motivated by the most trivial of applications. Although computer games are intended purely as entertainment, they are, nevertheless, state-of-the-art in terms of technology.

Exactly the opposite experience had happened a few years earlier when colleagues asked me to attend a workshop about environmental modelling. Embarrassed that my model was still in an early stage of development, I flew in a few days beforehand and cobbled together a quick prototype that displayed colourful runtime graphics. My talk consisted entirely of running the model under different assumptions and conditions. Knowing how shallow the model was, I felt embarrassed even to present it. The scary thing was the audience's reaction. Instead of tearing my work to shreds, they were wildly enthusiastic. It turned out that because people could actually see what would happen, the model had enormous credibility. Several senior government officials in the audience wanted to put my model to work immediately in planning government policy. Perhaps wisely, I suggested instead that they talk to my colleagues who really knew what the problem was all about!

Anyone who plays a computer game takes it for granted that the game will provide near-perfect images, fast response time and a host of other features. Users simply will not tolerate a game that does not match up to the latest standards. If you have experienced virtual reality in the latest version of *Tomb Raider*, you will never go back to *Dungeons and Dragons* on a command line interface. Computer games are one of the biggest areas of the computing market. Their target market has driven many of the advances in computer graphics and simulation.

But there is more to this story than meets the eye. The simulation models that we use in research and management are just games but, unlike most other games, they are games with a serious purpose.

If a picture is worth a thousand words, then virtual reality is worth a million. There are many possible applications for virtual reality, over and above entertainment. One is to reconstruct ancient buildings from their ruins: rebuilding the Library of Alexandria or the Temples at Olympia not only helps historians and archaeologists in their research, but also makes the past come alive for all of us.

If virtual reality can make the past come alive, then it can also make it possible for us to pay virtual visits to remote and inaccessible places in the real world. Nature documentaries have been popular for decades. In coming years we will see virtual tourism (in some guise or another) appear as a cheaper alternative to the real thing. As we saw above, we could explore new places before we actually went there. Virtual tourism would also allow us to visit places most of us could never get to, such as the surface of Mars, or Arctic wilderness.

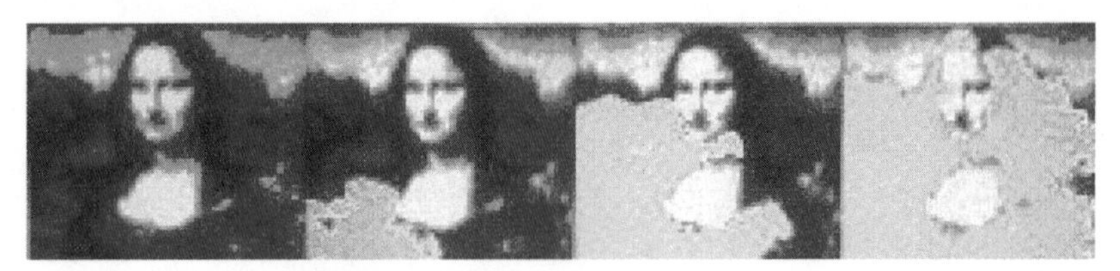

Figure 11.1 An experiment that you would never do in real life
The images show different stages in the simulated spread of a fire that starts in the bottom left corner. Notice that at the end the only remnant of the original picture is her smile!

A colleague of mine, Dirk Spennemann, designed a virtual field trip for his students. Most students are totally inexperienced at fieldwork, camping and planning outdoor activities. His virtual field trip helps them to avoid making costly, perhaps even dangerous mistakes out in the field. By playing this online game students gain valuable experience before they have to undertake real fieldwork later on. The game lets them see the consequences of their mistakes (e.g., failing to allow for poor weather), so that they can avoid making them in real time.

One of the virtues of models is that they allow you to do things that would be impossible in real life. For instance, it would be considered an act of sacrilege to destroy a great work of art. You could not get away with burning Leonardo da Vinci's portrait of the Mona Lisa, and yet, inside a computer, you can do things like this again and again with complete impunity (see Figure 11.1).

We can imagine many such experiments. Park managers cannot burn down a forest to see whether or not it will grow back. Civil authorities cannot spill toxic chemicals into a town's water supply to see what would happen. Nuclear scientists cannot melt down a nuclear reactor to see if safety precautions will work. Aeronautical engineers do not build a billion-dollar aircraft just to find out whether it can fly. People cannot perform any of these drastic experiments in real life, but they can inside a computer.

This is why virtual worlds have become such important tools in many professions. Pilots use flight simulators to train for emergency situations. Architects use virtual worlds to show clients what buildings will look like when they are finished. Doctors use virtual patients to try out and practise new surgical procedures. Engineers build virtual bridges to test their safety under extreme loads and high winds. Fisheries managers build models to ensure that fishing practices do not compromise the future of their industry.

Virtual worlds help people to cope with the complexity of the modern world. One of the hallmarks of complex systems is that they are inherently unpredictable. If your car is in a poor condition, and the weather is bad, it's late at night and you are in a hurry, then it's a fair assumption that you risk having an accident. However, you cannot predict exactly when, where or how that accident will happen. Instead, you can prepare for various scenarios. You can ask 'what if' questions and learn how to answer them. What if I slow the car down a bit? What if I keep extra distance between me and the next car? What if I avoid busy roads?

When it comes to the big issues, however, such as managing the world's environment, we do not have the luxury of trying experiments to see what approach will work best—for one thing, we do not have the time. Mature forests can take hundreds of years to regenerate. Nor can we risk destroying the very thing we are trying to preserve, but playing serious games with virtual worlds enables us to have our cake and eat it too. We can experiment with global warming. We can look at the implications of globalisation on the world's economy (see Chapter 12).

Can we build a world inside a computer?

Earlier we looked at the prospects of using virtual reality to recreate parts of the real world inside a computer. Just how feasible would it be to build a complete model of a real system, perhaps even of the Earth, inside a computer?

As Table 11.1 shows, the memory capacity of modern computers is now at a stage where it begins to compare favourably with the size of some natural systems. The data contained in the entire human genome, for instance, could be stored on the hard disk of many home computers.

These figures show that the data storage capacity of modern computers is now at a stage where it is meaningful to ask whether we should try to model complete systems in detail. However, simply having sufficient

Table 11.1 The number of components that make up various systems

System	*Number of units*
Australia	$\sim 8 \times 10^6$ square kilometres
Servers on the Internet	$\sim 1 \times 10^8$ computers
The Earth's surface	$\sim 5 \times 10^8$ square kilometres
Human genome	$\sim 3 \times 10^9$ base pairs
World population	$\sim 6 \times 10^9$ people
Human brain	$\sim 4 \times 10^{10}$ neurons
World's largest supercomputer	$\sim 1 \times 10^{11}$ bytes
Our galaxy	$\sim 1 \times 10^{11}$ stars
Human body	$\sim 1 \times 10^{13}$ cells
Storage capacity on the Internet	$\sim 1 \times 10^{18}$ bytes

memory capacity is not quite the same as being able to do it. There is still much that we do not know about each of these systems. Also, one single byte would not be enough to represent all the contents and features of a neuron in the brain. To represent all the connections involving even a single neuron would take many bytes.

Hans Moravec's book *Mind Children* contained the idea of simulating the entire Earth, along with every living thing, in minute detail.[4] Is this possible? The answer is that it depends on what you mean by 'minute detail'. If we take it to mean that we must simulate every single atom in the world, or even in one person's body, then the answer is most certainly 'no'. The body of a 100 kg person contains somewhere in the vicinity of 5×10^{27} atoms. That is 5 000 000 000 000 000 000 000 000 000 atoms. As Table 11.1 shows, this is more than a billion times the total number of bytes of storage on the Internet at the present time. In a few decades, if the increase in storage capacity continues at its present exponential rate, we may eventually approach this huge number. However, that does not mean that it would be possible to model every atom. For one thing, the computations would be horrendous. Even to calculate a simple change of state for such a vast number of items would take all the computers in the world about a year. But the real calculation would be hugely more complex, because it would involve computing the interaction of each atom with perhaps billions of neighbouring atoms. So even a single microsecond of activity would take our present computing resources millions of years to accomplish!

Some people will argue that quantum computers will eventually make such mind-boggling calculations feasible. But this misses the point. The real issue is that to capture the essential behaviour of a system we don't have to resort to modelling it in minute detail. As we saw in earlier chapters, all systems in the real world are modular. They are composed of units. Atoms organise themselves into molecules. In the human body, these molecules are mostly contained within cells, which are organised into tissues and organs. So molecules, cells and organs make up a hierarchy of modules. If we understand the way in which each module works, then we can ignore its inner workings when we try to model the entire system.

For instance, in trying to understand the spread of AIDS through a community, it is not necessary to model every detail of every person. All we need to model are the aspects of their behaviour that have the most direct bearing on the spread of the disease. We don't need to include such details as whether they have studied Ancient Greek, or rocket science.

The point is that we *can* build very useful and productive models. For instance, the Great Barrier Reef off the coast of North Queensland is actually not a single wall of coral at all. It consists of thousands of reefs spread out along a thousand miles of coastline. To understand the spread of the Crown of Thorns Starfish, a pest that destroys entire reefs, scientists built a model of the coast and all the reefs. This exercise did not require masses of detail, because the course of starfish outbreaks on each reef was well known and predictable. It was sufficient to model each reef as a simple point and to follow the flow of starfish larvae, as they spread from reef to reef like a nuclear chain reaction.[5]

Many organisations are putting these ideas into practice by modelling entire systems. Several cities have built models of their entire road networks, and all the traffic on them, so that they can study ways of ensuring traffic flows as quickly and smoothly as possible. In these kinds of models, each driver is represented as an 'agent' that behaves according to its own individual character. That does not mean that the scientists survey every person in the community. Instead, they set the agents up to react according to rules that reflect sociodemographic patterns. The crucial thing is that some drivers will head for the railway station and park there; others will take the freeway and park in town. Some will drop off children at school; some will take part in a car pool. Some will leave early, some later and so on.

By checking their models against observed traffic patterns, the modellers can check that they have enough of the right kind of details. They can then play with the model to see what sorts of problems emerge under unusual conditions, such as heavy rain or an accident on the freeway. They can also test the effectiveness of different road patterns, such as a new set of traffic lights or a bypass.

Is seeing believing?

In the year 1972, the United States government was thrown into disarray by the Watergate scandal. President Richard Nixon and his staff were accused of covering up their involvement in a break-in to the Democratic Party headquarters at the Watergate hotel. The controversy centred on tapes that Nixon had made of conversations with his staff in the Oval Office. Significant sections of the recordings were missing, leading to suspicions that crucial passages had been deleted to remove incriminating dialogue. Experts were called in to examine the tapes and to look for evidence of tampering, but their results were inconclusive.

In the early 1970s, the technology for manipulating recordings was pretty crude. Pressing buttons on a tape recorder, for instance, left telltale traces in the signal on a tape. Sound editors in studios would literally cut and splice pieces of tape. Technology has come a long way since then. Today, there is any number of advanced systems for seamlessly combining different sounds, as well as for manipulating images.

We can see some of the most dramatic evidence of these advances in movies. Instead of relying solely on elaborate sets and special effects, actors' images can be transported into any background that the director wants. Editors can add or remove characters from a scene, a technique that has been used for many years by movie makers. Usually, however, the effects were confined to static elements, such as an image of a city on the horizon. However, editors can now paste moving figures into a scene. Among the best examples of this are the dinosaurs in Steven Spielberg's movie *Jurassic Park*. Frighteningly real on the screen, most of them were just images that had been pasted into scenes. By 2001, the Japanese film *Final Fantasy* was able to include characters and scenery that looked convincingly real, despite being entirely animated. Real people can now be animated, so long-dead actors can appear in new roles. Given the astronomical salaries of top performers, some suggest that the studios might do away with living actors altogether.

Is seeing believing? Given modern technology's ability to simulate real life so convincingly, can we really be sure that the images we see on the screen are real? This capacity to alter images and sound at will raises some serious questions. Anyone with a home computer and the right software could, for example, doctor photographs of a crime scene. They could replace the face of the real criminal with that of anyone they wish. Admittedly, this is perhaps unlikely, but the possibility raises the very real problem that real criminals caught on tape could argue that the evidence had been altered. There are other unpleasant prospects as well. Innocent people could be cut and pasted into compromising photographs which, although they may be easily disproved, could be used in smear campaigns or in other ways to damage reputations or manipulate public opinion. How can we ever be sure that any image is genuine?

If it is possible to fake anything, then it is also impossible to prove anything. The plot of the movie *Wag the Dog*, centred on the idea of a government supporting its political agenda by falsifying evidence for the media. Real-life replicated fiction when Middle East countries cried out for proof that Osama bin Laden was responsible for the September 11 attacks on the United States. When the United States finally released a video that showed bin Laden actually admitting his involvement, his supporters immediately claimed that the video had been faked. This incident highlights a general problem: the potential to fake evidence provides an opportunity for people to reject any evidence that contradicts their prejudices. Extreme groups or individuals can argue 'conspiracy' against anything that contradicts what they wish to believe. Thus some people persist in claiming that the Holocaust never happened, that man never landed on the moon, or that the US government is holding aliens prisoner in Area 51.

Are eye witnesses reliable?

Everyone has seen it. The 'pea and shell' game is as old as history. The magician places the pea under one of the shells, then swaps the shells around and around while you watch. At the end, you pick out the shell where the pea should be, but it's not there. This is a classic example of sleight of hand. What we fail to see is that in the course of moving the shells, the magician palms the pea, transferring it from one shell to another. Sleight of hand is a good example of the way in which our senses and perceptions of events can be deceived.

If our perceptions can be fooled, so too can our memories. Our perception of reality is a fragile thing and this is especially true of our memories. For instance, there are places that I can remember visiting, and yet I can't tell for sure whether I have really been there or just visited in particularly vivid dreams. Likewise, memories of places that we've visited in cyberworlds can be just as strong, just as 'real', as the real world. Of course, this raises severe concerns about the reliability of eye witness reports in criminal cases.

Efforts to weed out child abuse, for instance, have been plagued by instances in which those questioning the children have unwittingly implanted false memories. In their book *The Myth of Repressed Memory: False Memories and Allegations of Sexual Abuse*, Elizabeth Loftus and Katherine Ketcham describe several cases in which psychiatrists had unwittingly planted false memories of childhood sexual abuse in patients who were under hypnosis.[6] In one case, guidance from a therapist helped a girl to remember that her clergyman father had regularly raped her between the ages of seven and fourteen. When the allegations became public, the father was forced to resign his position. A medical exam, however, established that she was still a virgin.

One reason for incidents of this kind is the very nature of perception. Every day an incredible barrage of data assaults our senses. The mind cannot remember it all—even the supreme parallel processing equipment in our brains cannot process everything. The only way we can cope with this torrent of incoming data is to selectively ignore most of what we see, hear and feel. Daniel Dennett has argued that what we experience as 'consciousness' is based on relatively few key features about the world we experience.[7] The most telling of these features are changes. Dennett's claim was controversial when it was first put forward, but subsequent research has since confirmed how little we really do process.

Psychologists suggest that dreams are the mind's way of sorting out our experiences, of making sense of the world around us. From time to time, most of us have vivid dreams, in which it seems that we enter other worlds. Sometimes, these dreams seem to take people back, time after time, to the same place. From time immemorial, we have sought ways to recreate our dream experiences in our waking hours. The Australian Aboriginal peoples, for instance, talk of the Dreamtime. Likewise, every culture has its spirit worlds, and a rich array of legends and stories to go with them.

In the modern world, plays, novels, movies and TV sitcoms all take their audience into different worlds. They take them to worlds where their everyday worries disappear, where anything is possible. If they are timid, they can be bold. If they are poor, they can be rich. To some people, the virtual worlds of make-believe are more real than the real but dreary world in which they live. The line between fiction and reality sometimes blurs. Millions are addicted to soap operas. When a character dies, or gets married in a popular TV series, it is almost a national event. Actors and actresses inhabit a magical realm that brings the storybook to life.

In computer games, our dreams can take on tangible form. In a movie, you can watch characters play out their roles in outer space or ancient Rome, but the experience is entirely passive. With a computer game, on the other hand, you become one of the characters. The whole course of the story can take many different turns, and the outcome depends on what you do.

Cyberspace

In his book *Neuromancer*, William Gibson introduced the term *cyberspace* to refer to a virtual universe of information and ideas.[8] The term derives from Norbert Wiener's concept of *cybernetics*, the study of how systems organise, reproduce, evolve and learn. Gibson's world not only contained facts and information, but was inhabited by intelligent beings as well. Thanks to the Internet, cyberspace has now taken on a tangible form. As well as the real world that we inhabit, we can also enter cyberspace via the Internet and move from site to site, exploring facts and information. We can communicate with other users, not only by email, but also by entering chat rooms, playing interactive games online, or leaving messages on notice boards.

For some people, their activities in cyberspace can be more real than their lives in the real world. In this sense, cyberspace's effect on people is no different from that of many other alternative worlds: the adventure novel you can't put down, the mystical world of transcendental meditation, or the long-running TV soapie you watch every evening. However, cyberspace does differ from these other universes in two important respects. First, you can interact with it. Second, in cyberspace you can also interact with other people. Role-playing is an important attraction of cyberspace. For a few hours at a time, you kiss your mundane existence goodbye.

Many science fiction writers have speculated on where this escapism might lead. In the movie *Total Recall*, Arnold Swarzenegger's character is offered a virtual holiday on Mars. In *Star Trek*, the star ship *Enterprise* has a 'holodeck', where the crew can relax by visiting virtual worlds of their own choosing. In *The Matrix*, the human population live out their lives plugged into a giant computer that simulates an artificial world for them. They are never aware that their real bodies are floating in nutrient baths inside a vast storage chamber.

Stories such as these are disturbing because they force you to question the very nature of reality. Modern travel tends to blur reality too. If you walk from A to B, then you experience the entire journey as a continuous flow of reality. Modern travel restricts the sensation of travel, making the experience less real. When I was a small child, I used to amuse myself on long car trips by imagining that the scenery outside was created around the car. Because I could not reach out and touch it, it existed only as the images I could see. So, like Jim Carrey's character in *The Truman Show*, I pretended that the places we visited were special sets created for my benefit. The scenery passing us by *en route* from one place to another were simply images created to make the journey seem real.

During plane travel, the sensation of reality during travel is dimmed even more. You get on board at A, which is a real place. You sit in a capsule eating and watching movies for a few hours. Then you leave again and find yourself in a new world called B. You have no sense of continuity between the two worlds.

In Aldous Huxley's novel *Island*, the author includes a mynah bird that regularly chimes in with the phrase 'Here and now boys' to pull characters back to reality.[9] Psychiatry, not to mention many philosophies and religions, stresses that the only place that is real is here and now. One of the hallmarks of many kinds of mental illness is the inability to distinguish reality from fantasy. All of us live our lives partly in this world and partly somewhere else. We spend a lot of our lives playing out roles. People wear suits to play at being successful professionals; young men drive hot cars to play the macho male, and so on.

In cyberspace, we can spread our wings and adopt any persona that we want. We can become anything or anyone, from a dinosaur to our favourite action hero.

The virtual society

In her book *The Pearly Gates of Cyberspace*, Margaret Wertheim describes how the medieval idea of heaven, and other spiritual worlds, has been transmuted in the electronic era into the notion of cyberspace.[10] As we have seen, the Internet takes cyberspace into whole new realms, but the jump to real-time communications was motivated initially by game players. *Dungeons and Dragons* is a popular computer game in which the player has to navigate his or her way through confusing corridors, dodge lurking dangers and collect various treasures. At first it was purely text-driven, and full of instructions such as 'You are in a maze of twisting passages all alike'. Because it required only the transfer of text, it was relatively easy to put the game online. This made it possible for many users to participate in the same game, all at the same time. The result was *Multi-User Dungeons and Dragons*, or MUDDs for short.

It was immediately apparent that the protocol used for playing games could also be used for many other purposes. Internet Relay Chat (IRC) allowed groups of individuals to carry on conversations. Chat rooms became a popular way of socialising and 'meeting' people with similar interests. In chat rooms, participants often take on pseudonyms, often reflecting who they 'wanna be', rather than who they really are.

Chat rooms are extremely addictive. Some people even prefer online discussions to direct conversations. For example, students in a computer lab are perfectly able to talk to each other. And yet, during lab sessions they will carry on lively conversations with each other via chat rooms and email, even when the person they are conversing with is sitting in a chair right next to them!

Perhaps the most telling innovation has been to link virtual reality with communications technology. We can enter fictional worlds and play active roles in them through our avatars (see page 161). Unlike earlier protocols, however, the virtual world is portrayed on the computer screen, so we can see where we are and what is around us. Not only can we move around and do things, but we can interact with avatars controlled by other people. And those people could be anywhere in the world.

There are many potential applications for online virtual worlds. For instance, one application for this technology that has immediately been adopted is 'interactive games'. Kids can join a commando team, composed of recruits from all over the country, and undertake a mission against a similar team of opponents. However, there are other, more serious applications. International meetings, for instance, are extremely expensive to stage. With dozens of conferences each year in any given profession, it is physically impossible to attend them all. A viable alternative is to participate via cyberspace. At a recent national conference, two of the keynote speakers were unable to attend in person, but gave their speeches online. This is increasingly common—there are even conferences where no-one turns up in person at all. These virtual conferences are held completely online. Authors submit their papers in the usual way, and discussion about the papers takes place via email and chat rooms.

chapter 12

THE GLOBAL VILLAGE

One of the problems the Internet has introduced is that in the electronic village all the village idiots have Internet access.

Peter Nelson[1]

Rare and momentous events such as the turn of a millennium have always been the occasion for prophecies of doom. It is typical of the information age that these predictions of doom now centre on computers. In the years immediately leading up to the year 2000, the media made a huge fuss over the 'Millennium Bug', a supposedly fatal flaw at the heart of computers that would trigger the collapse of the entire socioeconomic framework. No such collapse occurred.

Nevertheless, the world is beset by huge problems. We are perpetually confronted by threats such as climate change, crime, epidemics, famine, globalisation, injustice, intolerance, nuclear and biological warfare, overpopulation, pollution, poverty, species extinctions, social breakdown and terrorism, to name but a few. Never before have people faced the vast and rapid changes that we do today. Never before have technological changes come in such profusion, at such speed and with such vast, immediate and unpredictable consequences. Information technology is at the heart of many of these changes. It is serendipity gone mad.

Our brains are geared to simplify. To cope with the enormous volumes of information that we encounter in the modern world, we respond by ignoring most of it. We tend to deal with the familiar, and look for simple solutions. In his book *Future Shock*, Alvin Toffler vividly described the problems that people have in trying to cope with the pace of change in the modern world.[2] The future overwhelms us with its complexity.

Computers and information technology can help us solve most of these problems. Global communications and vast data storage warehouses help people make informed decisions. They can make us aware of the consequences of our actions and, as we have seen in earlier chapters, they have the potential to help us make sense of the complexity that besets the modern world.

But, what we often get are 'solutions', not answers. Computer vendors talk about solutions, when what they really mean are software and hardware products. The implication is that if you buy this or that piece of equipment, it will solve your problems. However, as we have seen, the unexpected effects of computers, and the information they contain, on our lives often create just as many problems as they solve.

Global incorporated

> *Computer technology is altering the form, nature, and future course of the American economy, increasing the flow of products, creating entirely new products and services, altering the way firms respond to demand, and launching an information highway that is leading to the globalisation of product and financial markets.*
>
> US Bureau of Labor Statistics[3]

Information technology is just one of the major underlying forces enabling the rapid trend towards globalisation. The first, and most obvious, is communication. In the past, empires rose and fell on the effectiveness of their communication networks. It is said that when Ghenghis Khan died, the news of his passing took years to disseminate throughout his vast Mongol empire. When Lady Diana died, a worldwide audience watched her funeral live on television. Between these extremes lie vast changes in the way we conduct our lives.

Global business is possible because companies on opposite sides of the world can talk to each other. For instance, advanced communications enable us to exchange money anywhere in the world. This, in turn, enables people to carry out electronic commerce around the globe. Perhaps the most telling results are the establishment of international money markets, and the opening up of stock exchanges around the world to international traders.

The need for on-the-spot, up-to-date information is not confined to large organisations. To take an example: suppose that a young married couple living in Edmonton, Canada look to investments as a way of boosting their savings and securing their future. They look at prospects not only within Canada, but also around the globe. In the evening they go online and check out the stock exchanges in Australia, Tokyo and Hong Kong. In the morning they do the same for London and Bonn. If they find companies they are interested in, they naturally want to find out more. So they might be looking at such widely spread prospects as a tour company based in Dunedin, New Zealand; a chain of micro-breweries in Portland, Oregon, or a company building intelligent robots in Edinburgh, Scotland. In each case they are able to access detailed local information, not only about the company, but also about the area, local competition, and so forth.

Advances in communication have changed the playing field on which business operates. Given their imperative to grow, companies have to exploit communications to find new markets and to become more efficient. This has two apparently contradictory effects. On the one hand, it enables large corporations to conduct their business on a larger scale than ever before. On the other, by promoting and selling via the Internet, small, specialised businesses are flourishing as never before.

The technology tax

For the past several generations, information technology has been making rapid changes. These changes, and especially the increasing rate at which they are occuring, have had a powerful impact on our society.

Consider music. After Edison invented the phonograph, his original cylinders were soon replaced by vinyl records, which became the standard for music reproduction for most of the twentieth century. For decades, magnetic recording tape failed to catch on as a medium because it was awkward to use, until the invention of the tape cassette led to an explosion of tape recording. Cassettes were more compact than records, but had one great sale-destroying disadvantage. You had to wind through them to get to the track that you wanted. Records, in contrast, were a random access technology. You could jump instantly to any track that you wanted to hear. However, perhaps the decisive factor was that, right from the start, cassette decks included recording heads in their mechanism. As a medium, therefore, cassettes had the advantage that you could buy blanks and record your own music.

These two technologies, cassettes and records, co-existed for an entire generation, until compact disks (CDs) came on the scene. Their important innovation was digital recording. Both records and cassette tapes used an analogue system. In analogue recording, the waveforms in the sound are represented on tape by continuous changes in the magnetic signal. The problem with this approach is that it was too susceptible to noise. Wear and tear on records soon produced background hisses and crackles that marred the sound. These faults were somewhat reduced, but not eliminated, by the invention of Dolby filtering. Digital recording, on the other hand, produces no background noise at all. The waveforms are sampled many times per second and converted to numbers that represent the volume of different frequencies. These numbers are stored on the CD as a series of binary dots burned into the surface. The advent of the MP3 standard for digital media is now leading to a new revolution. Some artists are bypassing recording companies and releasing their works directly on the Web.

Similar changes occurred in the recording of video, but on a shorter time scale. For decades celluloid film was essentially the only medium for film recording. Video cassettes opened the market for home movies in the 1980s but, by 2000, were giving way to DVDs. Other technologies, such as online, on-demand movies, are just around the corner.

What these stories show is the steadily accelerating pace at which new information technologies take hold of society. The introduction of any new technology follows a pattern consisting of three stages. In the early stages, the uptake of the technology is slow, and largely confined to 'early adopters'—affluent and often technically savvy individuals, who can afford to take a risk with something new and unproven. When the number of users reaches a critical level (about 10–15 per cent of the population), the rate of uptake increases explosively. Finally, as the number of users approaches saturation, the rate of uptake slows down again as only small pockets of non-users remain.

There are several reasons for the explosive growth phase. One is that most people need time to become comfortable with a new technology. Do the neighbours have one? Does it work OK? There is also the question of price. New products are expensive to produce and these costs get passed on to the customer. As demand increases, economies of scale become possible so the price goes down and demand goes up.

The time scale on which all this happens can vary enormously. For the automobile the process took decades. Today, companies with new products usually plan on the process taking a few years at most. Companies can strike trouble at each stage. If they fail to crank up production and efficiency at the onset of the explosive stage, then they are likely to be driven out of business by competitors. Likewise, many industries have trouble adapting when they saturate the market and cannot scale down production or find new markets.

In communications, especially in computing, the normal process is confused because research is constantly developing new technologies to replace the existing ones. Before a new technology has saturated the marketplace, a new one has appeared and begins to replace it.

My own case is probably typical. Although I have used computers at work for over 30 years, at home I was by no means an early adopter. My family bought its first personal computer in 1985. Since then, however, we have bought a new one every two-and-a-half years on average. These purchases were driven by 'necessity': the need for more disk space, for new software, for multimedia, Internet access and portability.

If the changes in hardware have been fast, the changes in software have been even more rapid. Companies constantly compete to come up with better features to make you buy their product. This means that we are constantly being deluged with new products. In theory, there is nothing to stop you from saying 'Enough is enough. I've got my computer. It works perfectly well. I see no need to upgrade.' Some people and some companies get away with doing this. At least for a while. But they are the minority.

Two factors make it impossible to stand still. The first is maintenance. As the new hardware improves, the older hardware gets discarded. Even if you can get your faulty disk or power supply fixed, it is probably cheaper to upgrade. Once you do that you find that many of your old programs no longer work under the new operating system. So you end up having to buy a new suite of software as well.

The second factor is peer pressure. If your business is self-contained, it may not matter if your software is out of date. But large organisations send their suppliers and clients electronic files all the time. If you stubbornly try to remain fixed, then you soon find that people are sending you crucial information that you cannot access with your out-of-date systems. Similarly, if you stand still you are likely to find that your colleagues or, worse still, competitors are using new technologies such as multimedia or the Internet, and leaving you behind.

The spread of the Internet means that you can no longer be an island, complete unto yourself. Internet developers are always keen to present the best online image they can. This means having the glossiest designs, and all the latest whiz-bang features. In using the latest features they assume that other users have the latest versions of Web browsers and other applications. These software packages in turn require the latest hardware and operating systems. So when an elderly aunt sends you an email attachment that your ancient (translation—more than twelve months old!) system cannot even open, then you know it's time to upgrade.

Maintenance and peer pressure force individuals and organisations to climb aboard a technological treadmill that is difficult to get off. Of course, there is a price. Upgrading hardware and software at regular intervals means that you have to spend hundreds, possibly thousands, of dollars every year just to keep up. This cost has been so widespread and so consistent since the introduction of personal computers, that it is not appropriate to think of it as a one-off start-up cost. Rather, it is an ongoing annual expense. Think of it as a technology tax. It is the price of entry into the modern professional workplace, but it is much more than that. For many companies, it is the price they must pay for survival in an ever more competitive market.

The global village

It is evening in Kuala Lumpur and I have just returned to my hotel room. I turn on the television to see what it can show me about Malaysian culture. Instead, I find the Australian movie *Mad Max*. I flip to another channel. This time it is CNN's News with the latest on the US presidential election campaign. I flip to another channel. At last, there are some Malaysians on the screen. But the programme is a western-style quiz show.

Whether it is Brazil, Finland or Korea, the same story is repeated in country after country. With the television blasting an endless stream of western 'culture' at the younger generation, world culture is becoming increasingly homogeneous. And it is not just what you can see on TV. The homogenisation of culture extends to dress, to shopping and restaurants and, most importantly, to the way people live, think and behave.

Marshal McLuhan famously described the results of modern communications as a 'global village'.[4] His argument was that communications are now so good, so universal, that we are all part of a single, interconnected community. In many respects that claim is now literally true.

If films and television have made a huge impact on culture and values world wide, then the Internet is likely to have just as great an impact. For one thing, English is the dominant language on the Internet. At first, all Websites were in English. Other languages only started appearing as the user base expanded. However, many of those sites are actually bilingual. Often the first page you enter will be in English, with an option to view pages in the local language. Web publishers are forced to do this because they know that if they want the rest of the world to read what they have to say, it had better be written in English. So all over the world, local languages, on the Web at least, are playing second fiddle to English. This bias puts a subtle, but real, shading on the way information is perceived. It is like a sport: if you want to be an international player, then you must play in English. Other languages are just for the small fry in minor, local leagues.

One effect of the dominance of English in all forms of media is that minority languages and cultures are dying out. According to the Endangered Languages Repository, there are around 7000 living languages in the world, but social pressures are driving many of them out of existence.[5] The Ethnologue catalogue classifies about 750 languages as extinct or nearly extinct.

Paradoxically, the Internet not only promotes the spread of English language and culture; it also helps threatened languages and cultures to survive. Isolation of native speakers from one another has been an important contributing factor in the demise of native languages. The Internet provides a medium in which speakers of a common language can communicate, and record details of their language and heritage. A colleague of mine, Dirk Spennemann, set up a Website for the culture of the Marshall Islands, a small nation consisting of islands in the Pacific Ocean.[6] The site not only contains language references but also traditional stories and legends and accounts of other aspects of culture, such as tattooing.

Polls and opinions

In the movie the *Rise and Rise of Michael Rimmer*, comedian Peter Cook plays the owner of an opinion poll company who rises to become dictator of Britain. One of the methods he uses to achieve his success is to poll everyone, thereby ensuring that his predictions are accurate. His election platform is to offer everyone true participation in the country's decision-making. When he takes office he decides every issue by running an opinion poll. Rimmer eventually becomes dictator because the population tires of completing opinion polls and votes to let him decide every issue for them.

In real life, parts of the Rimmer story are coming true. Opinion polls have played an increasingly important part in the decision-making of vote conscious western governments for many decades. Advances in telecommunications and information technology will see them having a greater impact still. In the United States, the cost of referenda has been so great that governments have been debating whether to decide some issues on the basis of sample surveys. At the time of writing, some governments have been toying with the idea of allowing voters to cast their ballots electronically via the Internet.[7]

This raises the serious question of whether we really need parliaments at all. Parliaments arose as a form of representation when time and distance prevented people from having their say, but in a global village is that still true? Many issues today seem to be decided by public opinion, but the danger in this is that opinion can be moulded. The mass media have enormous power to shape public attitudes by broadcasting the views that their owners want to promote.

At the same time that global communication is providing large organisations with greater social influence, the Internet is increasing the visibility and influence of small interest groups. Individuals who were once isolated from their like-minded fellows can now communicate with each other. This helps them to find an identity.

The Internet has already begun to play a major role in global politics. For instance, the rapid spread of Internet use in China became a dilemma for the Chinese government. On the one hand was the need to develop e-commerce and leading edge technology for the economic benefits. On the other, was the potential for widespread unfettered communication to upset the status quo.

During the 1990s, the Chinese government became concerned about the popularity of Falun Gong, a spiritual movement that combines callisthenics with Buddhist and Taoist beliefs. When it turned out that Falun Gong's followers were using the Internet to organise protests and other activities, the authorities closed down its Websites in China. However, the movement simply transferred information to sites in other countries, which initiated a struggle in cyberspace. The Chinese government set up a counter Website to discredit the organisation and installed filtering software to prevent access to outside sites. Followers of the movement claim that the Chinese government repeatedly tried to hack into its sites in North America and Britain. They also claimed that sites were attacked by overloading them. That it, the sites were drowned in automated connection requests that tied up their servers, so preventing access by other users.

There have been many other such incidents. During the conflict in Yugoslavia in 1999, Serbian-run Websites tried to counter NATO efforts to demonise them by reporting on the bombing from the Serbian side. In the days and months leading up to the resignation of Indonesia's President Suharto in May 1998, the Internet made possible the instant spread of anti-government information as well as enabling protesters to organise quickly, efficiently and in secret. During the lead up to the Iraq War of 2003, the anti-war movement made extensive use of the Internet to spread its message worldwide.

From cybercrime to cyberwar

The growth of the Internet as a universal vehicle for communication has opened a veritable Pandora's box of lurking problems and nasty practices.

In his novel, *The War of the Worlds*, H.G. Wells's Martian invaders meet their doom at the hands of humble earthly bacteria. The movie *Independence Day* brought this idea up to date by turning the bacteria into a computer virus that the humans upload into the alien computer network. Fiction though it is, this idea points to a new and insidious form of battle of the future—*cyberwar*.[8]

It is much cheaper, and no doubt a lot safer, to infiltrate enemy sites on the Internet than to mount a physical attack on their facilities. With e-commerce playing an ever more prominent role, the potential impact, and appeal, of cyberwarfare will increase to the point where it becomes inevitable.

There have already been numerous cases of hackers breaking into Websites dealing with politically sensitive issues and changing their contents in an embarrassing or damaging way. In 1997, for example, Portuguese hackers apparently broke into a Website run by the Indonesian Ministry of Foreign Affairs and loaded material that claimed the government had spread lies about events in East Timor. The Web was also one means of spreading information about the repression in the province prior to its independence.

In his book *The Lexus and the Olive Tree*, Thomas Friedman focuses on the struggle of local and indigenous peoples for greater autonomy. We can see this process in events such as the dissolution of the Soviet Union and in claims over land rights by native peoples in North America and

Australia. As the finances of large corporations take on the size and nature of national budgets, their disputes are likely to involve conflict of one kind or another. Other possible conflicts would be between factional groups, such as environmentalists, and large corporations.

The nature of these conflicts would be radically different from traditional warfare. Their increasing reliance on information and communications makes organisations vulnerable to digital warfare. The prominence of the Internet in commerce, government and defence leaves many countries susceptible to network terrorism. Recognising this, the US communication surveillance organisation, the National Security Agency, hired a group of 35 hackers during 1997 to simulate an attack on key Internet sites.

Another insidious development has been the use of the Internet to support terrorism. On 11 September 2001, terrorists crashed hijacked airliners into New York's World Trade Center, destroying both towers and killing over 3000 people. But as the world watched the twin towers collapsing in flames, few recognised the deeper, more sinister implications of the horror images. This attack signalled, in no uncertain terms, the globalisation of terrorism.

This dreadful event could not have taken place without long and meticulous planning, nor would it have been possible without excellent communications. Global information and communications have allowed terrorist groups in many different countries to cooperate, form networks and coordinate their activities. Instead of small groups being restricted to small attacks on local targets, they can now carry out large-scale attacks on targets of international significance. They are no longer limited by national borders or local resources. They can call on sympathisers for support wherever they may be.

Terrorism has always existed, but its increasing impact on the latter part of the twentieth century can probably be attributed to the power of modern communications and the influence of news media. Terrorism is effective only if people hear about it. One of the weaknesses of the media is that they report what they think will sell: sensation sells, and terrorism most definitely is sensational. If you want to get yourself and your cause on the evening news in every western country, then violence is the way to do it. One of the tragedies of terrorism is that it could probably be stopped, or at least greatly reduced, if the world's media decided to stop pandering to the terrorists and impose a blackout on reporting such incidents. But they have not and will not, for the simple reason that terrorism makes them money. Anyway if one organisation did not report it, then the next station or newspaper would.

Mass shootings were unheard of until a few deranged people decided to go out in style. But the media can never resist sensation and each new incident allows them to rehash the preceding horrors again and again and again, just in case we missed them the first time. So the first incident has spawned a string of copycat incidents. Now it is a seasonal event, not only in the United States, but in dozens of other countries, and the media report each and every one. Much the same happens with bomb threats, extortion attempts (especially if the extortioners threaten to poison popular food products) and other violent incidents.

The Internet adds a sick new dimension to this murder and mayhem. There are 'how-to' sites for terrorists, including bomb-making, terror tactics and recruiting new members.

In the increasing use of terrorism by some political groups we are perhaps seeing the birth of another new kind of warfare. Modern weapons make full-scale war between powerful nation-states increasingly devastating. A full-blown nuclear war would be so terrible that almost every government will do all it can to avoid it. What we are likely to see instead is conflict manifested in many other forms, such as disruption of communication and information resources.

Of nightingales and computer games

The song of the nightingale is beautiful. So beautiful that it became the favourite at the court of the Chinese emperor, according to Hans Christian Andersen's fairy story *The Nightingale*. When the bird is first brought to the palace, the emperor and his courtiers are entranced and can listen to its song all day. Then one day, the emperor receives an incredible gift—a mechanical nightingale. Not only does this artificial bird sing just as sweetly as the original, but it can sing all day and all night without ever growing weary. What is more, the mechanical nightingale is encrusted with jewels, not the dowdy brown feathers of the real bird, so it is beautiful to behold. Besotted with their new toy, the emperor and his courtiers soon lose interest in the real bird. Forgotten and neglected, the nightingale leaves the court and flies far away. Some time later, the mechanical bird breaks down. No-one can repair it. Stricken with sadness at his loss, the emperor falls ill. Desperate courtiers seek out the nightingale, realising that only its song can restore the emperor's will to live. At first, the nightingale is reluctant to return to the court where it had been shunned, but after much pleading by the courtiers, the bird finally agrees to return, but only for a single day each year. On hearing the bird's song again, the emperor recovers and the court never again takes the nightingale's singing for granted.

Like many children's tales, the story of the nightingale carries some important moral lessons. These lessons are even more apt in today's world than when the story was written. In a sense, today's 'global village' is a bit like that court in ancient China. In the real bird, we see represented all the wonders of nature. In the mechanical bird, we see the artificial world that technology, especially information technology (IT), has created. And, like the courtiers, we members of modern society are easily seduced by the magic of virtual worlds. In the process, we risk losing touch with our place in the real world. Just as the emperor fell ill when the mechanical bird broke down, so too are we endangered by our growing dependence on electronic gadgets.

Modern society is obsessed with technology, and with IT in particular. As we have seen, our ability to handle information is one of the keys to how well we cope with the complex and challenging world that we find ourselves in. At the same time, that very technology is itself reshaping society and producing problems of its own. The devastating effects that arose from the breakdown of the mechanical bird are just the sorts of complex, and unexpected, consequences that often accompany technological innovation.

From TV to the Internet, from mobile phones to laptop computers, IT is now an indispensable part of our daily lives. As our dependence on IT continues to grow, we find it changes the ways in which we do things. It even affects the way we think. We are constantly surrounded by hype about the latest gadgets, constantly being told that our quality of life depends on them. But it is important that we understand the nature of these changes, whether for good or ill, and where they are leading us. Will the serendipity machine help us to manage events in this brave new world of rapid change and global crises? Or has it already opened a Pandora's box and released global chaos and other unforeseen horrors?

FURTHER READING

Bossomaier, T.R.J. and Green, D.G. (1998). *Patterns in the Sand: Computers, Complexity and Life*. Allen & Unwin, Sydney.

Bossomaier, T.R.J. and Green, D.G. (2000). *Complex Systems*. Cambridge University Press, Cambridge.

Buckley, D.J., Ulbricht, C. and Berry, J. (1998). 'The virtual forest: advanced 3-D visualization techniques for forest management and research'. ESRI 1998 User Conference, 27–31 July 1998, San Diego, CA.

Endangered Languages Repository http://www.yourdictionary.com/elr/

Friedman, Thomas (1999). *The Lexus and the Olive Tree*. HarperCollins, London.

Green, D.G. and Bossomaier, T.R.J. (2001). *Online GIS and Spatial Metadata*. Taylor & Francis, London.

Koestler, A. (1972). *Then Roots of Coincidence—An Excursion into Parapsychology*. Random House, New York.

Kurzweil, R. (1999). *The Age of Spiritual Machines*. Allen & Unwin, Sydney.

Moravec, H. (1988). *Mind Children: The Future of Robot and Human Intelligence*. Harvard University Press, Cambridge, Mass.

Wertheim, M. (1999). *The Pearly Gates of Cyberspace*. Doubleday, New York.

NOTES

The Internet addresses cited here were all correct and functioning at the time of publication.

Chapter 1

1 See TOP500 Supercomputer Sites http://www.top500.org for details. A floating point number is any number such as 10.2 or 1234.5678. A floating point operation is any simple arithmetic step such as adding two numbers together. A Teraflop means a million million (1 000 000 000 000) floating point operations per second.
2 In computing, a byte corresponds to a single character. The following conventions are used to indicate the size of a computer file. Each term represents an increase in size by 10 powers of 2 (1024), which is roughly a 1000-fold increase.

Name	*Power of 2*	*Number of bytes*
Byte	1	1
Kilobyte	10	1 024
Megabyte	20	1 048 576
Gigabyte	30	1 073 741 824
Terabyte	40	1 099 511 627 776
Petabyte	50	1 125 899 906 942 624

3 You can find the Internet Software Consortium at http://www.isc.org.
4 See S. Lawrence and L. Giles (1999). 'Accessibility and distribution of information on the Web'. *Nature* 400, pp. 107–9. Also see http://www.metrics.com.
5 The Google Search Engine is at http://www.google.com.

Chapter 2

1 The Environmental Resources Information Network (ERIN) commenced operations in 1990. The environmental information systems that it set up became a model for similar projects in many other countries. The systems later became absorbed into the online biodiversity and environmental services provided by Environment Australia (http://www.environment.gov.au).

2 See R.J. Hnatiuk (1990). *Census of Australian Vascular Plants.* Australian Flora and Fauna Series No. 11. Australian Government Printing Service, Canberra.

3 The number of combinations C is given by the well-known formula $C(N,n)=N!/(n!(N-n)!)$, where N is the size of the pool and n is the size of the groups you are drawing from the pool. The exclamation mark '!' means a factorial product. For instance, $6! = 6 \times 5 \times 4 \times 3 \times 2 \times 1 = 720$. So $C(6,2) = 6!/(2! \times 4!) = 15$.

4 Suppose that there are 30 people at the party. The chance of any particular pair of people sharing a birthday is 1/365. In a group of 30 people, there are 435 different pairs of people, so the chances of no pair sharing a birthday is $364/365 \times 364/365 \times 364/365 \times \ldots$ repeated 435 times! This number turns out to be approximately 0.303. This means that the chance of having no birthdays in common is 30.3 per cent. But this means that the chance of some pair sharing a birthday is 69.7 per cent. In other words, the chances are good.

5 In general there are $N!$ permutations (arrangements) of any group of N items. For instance, for A,B,C, there are 3! or 6 orders: ABC, ACB, BCA, BAC, CAB, CBA.

Chapter 3

1 From *Walden—Or Life In The Woods* (1854) by Henry David Thoreau (1817–1862) [Project Gutenberg Etext #205, 1995].

2 At a certain point, the program had to convert a 64-bit floating-point number into a 16-bit signed integer. There is a potential trap in such a conversion: A floating-point number (the first format) can be as large as you like, but an integer (the second format) must lie within the range –32 768 to +32 767. Any number outside the allowable range would cause an overflow.

3 Machine code tells the computer what to do at the most basic level of hardware. A typical command might read something like 'Move the number in register 001 to register 002'.

4 From *The Art of War* by Sun Tzu, Chapter VI, paragraph 14. Translated and edited by Lionel Giles (1910) [Project Gutenberg Etext #132, 1994].

5 For a well-balanced binary tree the average path length between nodes is $\log_2 N$, where N is the number of nodes. For example, in a network of (say) 1000 computers, direct wiring to connect each machine to every other machine would require 500 000 lines. In contrast, a binary tree could link all of the machines using just 999 lines, but data transfer between two machines would require hops between up to 18 intermediate machines. By adding more than two links at each node, however, the maximum path required rapidly diminishes.

Chapter 4

1 Paul Claudel (1868–1955), quoted in P.H. Caskell (ed.), *Structure of Non-crystalline Materials* (1977). Taylor & Francis, London.
2 From the poem 'To a mouse, on turning up her nest with a plough, November 1785' by Robert Burns, in *The Project Gutenberg Etext of Poems and Songs of Robert Burns.* [Project Gutenberg Etext #1279, 1998].
3 See Stephen Jay Gould (1989). *Wonderful Life: The Burgess Shale and the Nature of History*. Penguin, London.
4 Kanchit Pianuan, Mingsarn Santikarn Kaosa-ard and Piyanuch Pienchob (1994). 'Bangkok traffic congestion: is there a solution?' A translation of the fifth issue of TDRI's White Paper Series (publication code: WB5), editor: Linda M. Pfotenhauer, in *TDRI Quarterly Review* 9(2), pp. 20–3. http://www.info.tdri.or.th/library/quarterly/text/traffic.htm.
5 See Edward Tenner (1996). *Why Things Bite Back: Technology and the Revenge of Unintended Consequences.* Vintage Books, New York.
6 I described this in detail in T.R.J. Bossomaier and D.G. Green (1998). *Patterns in the Sand: Computers, Complexity and Life.* Allen & Unwin, Sydney.
7 Gleick, J. (1999). *Faster: The Acceleration of just about Everything.* Little, Brown & Co., London.
8 The Blackout History Project. http://www.blackout.gmu.edu/highlights/blackout65m.html

Chapter 5

1 Edward Kasner and James Newman (1940). *Mathematics and the Imagination.* Penguin, London.
2 See T.R.J. Bossomaier and D.G. Green (1998). *Patterns in the Sand: Computers, Complexity and Life.* Allen & Unwin.
3 See the TOP500 Supercomputer Sites at http://www.top500.org.

4 One benchmark used to rank the speed of supercomputers is the Linpack problem, introduced by Jack Dongarra, which requires a computer to solve a dense system of linear equations. By varying the size of the problem (the number of variables and equations), and applying a standard algorithm to find the solution, you can get a good comparison of the maximal performance of different computers. The Linpack Benchmark is described in full at http://www.top500.org/lists/linpack.html.
5 Arthur Marcel (1998). 'Trouble with the metric system'. A talk broadcast on ABC Radio's *Ockham's Razor*, program, 2/8/1998.
6 The Beowolf Project (clustering PCs into parallel supercomputers) http://www.beowulf.org.
7 See the TOP 500 Supercomputer Sites at http://www.top500.org.
8 SETI itself featured in Carl Sagan's novel *Contact*. Century Hutchinson, London.
9 SETI@home online (public participation in the search for radio signals from extraterrestrial civilisations) is at http://setiathome.ssl.berkeley.edu.
10 You can find PiHex (project to determine the value of Pi) http://www.cecm.sfu.ca/projects/pihex/pihex.html.
11 Casino-21 (distributed modelling of future world climate) is at http://www.climate-dynamics.rl.ac.uk.
12 Folding@home (computing shapes of large proteins) is at http://www.stanford.edu/group/pandegroup/Cosm.
13 The Golem@Home project (evolving robot designs) site is at http://golem03.cs-i.brandeis.edu/download.html.
14 Find the Globus project (grid computing) at http://www.globus.org/about/default.asp.
15 See Eric Drexler (1986). *Engines of Creation.* Anchor Press, New York.

Chapter 6

1 Carl Correns in Germany, Hugo de Vries in Holland and Erick von Tschermk in Austria.
2 The *Dublin Core* is at http://purl.oclc.org/metadata/dublin_core.
3 See S. Lawrence and L. Giles (1999). 'Accessibility and distribution of information on the Web'. *Nature* 400, pp. 107–9. http://www.metrics.com.
4 For more details about XML and XSL refer to the World Wide Web Consortium http://www.w3c.org.
5 For more details about MathML, MusicML, ChemML see W3C http://www.w3c.org.
6 See D.G. Green, and T.R.J. Bossomaier (2001). *Online GIS and Spatial Metadata.* Taylor & Francis, London.

7 Species 2000 is at http://www.species2000.org.
8 Prominent standards for specifying these kinds of services include CORBA and the W3C's Resource Description Framework (RDF). For details refer to the W3C at http://www.w3c.org. A significant geographic application is the OpenGIS specification (see http://www.opengis.org).

Chapter 7

1 The program ELIZA was the product of Joseph Weizenbaum's research into artificial intelligence at MIT in the 1960s. Many interactive versions are available online. For the original description, see J. Weizenbaum, (1966). 'ELIZA—a computer program for the study of natural language communication between man and machine'. *Communications of the Association for Computing Machinery* 9(1), pp. 36–45.
2 P. Hogeweg and B. Hesper (1983). 'The ontogeny of the interaction structure in bumblebee colonies: a MIRROR model'. *Behavioral Ecology and Sociobiology* 12, pp. 271–83.
3 For more about Knowledge Query Markup Language (KQML) see http://www.cs.umbc.edu/kqml.
4 See Mark Rubinstein (1998). 'Comments on the stock market crash: eleven years later'. In the Society of Actuaries, *Risks in Accumulation Products*. Monday 19 October.
5 In many cities, security cameras (public and private) are becoming increasingly common, especially around bars, nightclubs, railway stations, major intersections and commercial premises.
6 See www.fipr.org/rip.

Chapter 8

1 R.A. Brooks (1986). *Achieving Artificial Intelligence Through Building Robots*. AI Memo 899, MIT, Cambridge, Massachusetts. R.A. Brooks (1991). 'Intelligence without representation'. *Artificial Intelligence* 47, pp. 139–59.
2 T.S. Ray (1991). 'An approach to the synthesis of life', in C. Langton, C. Taylor, J.D. Farmer and S. Rasmussen (eds). *Artificial Life II*. Santa Fe Institute Studies in the Science of Complexity, vol. 11, pp. 371–408. Addison-Wesley, Redwood City, California.
3 Evolutionary computing (EC) is a generic term for a large collection of methods which go by many names, mostly based on differences in their approach.

Some of these, such as genetic algorithms and evolutionary programming, are described briefly in this section.

4 See J.H. Holland (1975). *Adaptation in Natural and Artificial Systems.* University of Michigan Press, Ann Arbor.

5 W.J. Freeman, (1992). 'Tutorial on neurobiology: from single neurons to brain chaos'. *International Journal of Bifurcation and Chaos* 2(3), pp. 451–82.

6 See Richard Dawkins (1986). *The Blind Watchmaker.* Longmans (republished in 1998 by Penguin Books), London.

7 Cambrian Art, developed by Mattias Fagerlund was to be found at the site http://www.cambrianart.com. The service became inactive early in 2003. However, a search of the Web is likely to turn up other online demonstrations of evolutionary design.

8 For example, two common assumptions are that processes are linear and that the effects of different variables are independent of one another.

Chapter 9

1 Deoxyribose Nucleic Acid (DNA). The double helix model was proposed in 1954 by two brash young biologists, James Watson and Francis Crick, working at Cambridge University. They based their model on data gathered by their colleagues Maurice Wilkins and Rosalind Franklin. Watson vividly recounts those events in his famous autobiography account, *The Double Helix* (first published in 1968). In a more recent account, *Rosalind Franklin: The Dark Lady of DNA* (HarperCollins, 2002), Brenda Maddox recounts the crucial contributions that experimental data played in the discovery.

2 Bioinformation is biological information, but the term is usually used in the context of genes and proteins. Bioinformatics is the science of gathering and interpreting bioinformation.

3 RNA (Ribonucleic Acid) is a long-chain molecule that copies the genetic code and transfers it to where it is used. The RNA sequence differs in two main ways from DNA. There is just a single chain, and the fourth base in the RNA code is Uracil instead of Thymine. RNA appears in several forms (e.g., messenger RNA or transfer RNA).

4 Ribosomes are bead-like structures composed of units of protein and RNA. Located in the endoplasmic reticulum, they are the cell's protein-making factories.

5 See P. Onyango, W. Miller, J. Lehoczky, C.T. Leung, B. Birren, S. Wheelan, K. Dewar and A.P. Feinberg (2000). 'Sequence and comparative analysis of the mouse 1 megabase region orthologous to the human 11p15 imprinted domain'. *Genome Research*, 10, pp. 1697–1710.

6 Prosite http://expasy.hcuge.ch/sprot/prosite.html.
7 The method was first described in S. Fodor, J.L. Read, M.C. Pirrung, L. Stryer, A.T. Lu and D. Solas (1991). 'Light-directed, spatially addressable, parallel chemical synthesis'. *Science* 251, pp. 767–73.
8 Fodor *et al.* have built arrays with up to 400 000 cells.
9 GenBank http://www.genbank.org.
10 DNA Database of Japan (DDBJ) http://www.nig.ac.jp.
11 European Molecular Biology Laboratory http://www.embl-heidelberg.de.
12 One of the earliest large-scale studies is described in W.R. Taylor, (1990). 'Hierarchical methods to align large numbers of biological sequences'. R.F. Doolittle (ed.), *Molecular Evolution: Computer Analysis of Protein and Nucleic Acid Sequences. Methods in Enzymology* vol. 183. Academic Press, New York, pp. 456–74.
13 The Human Genome Project http://www.ornl.gov/hgmis/faq.
14 To reassemble a sequence from fragments, a computer determines which strands overlap with one another. This task is more difficult than it sounds. For a start, you have to get the alignment between any pair of sequences just right. If each strand is (say) a thousand bases in length, and assuming there is an overlap, then there are 3998 possible ways they could match. But if you have (say) 100 000 fragments, then there are 100 million pairs to be checked. This means you need to consider about 40 trillion (4×10^{12}) possible ways in which strings might match each other.
15 See F. Jacob and J. Monod (1961). 'Genetic regulatory mechanisms in the synthesis of proteins'. *Journal of Molecular Biology* 3, pp. 318–56.
16 G. Halder, P. Callerts and W.J. Gehring (1995). 'Induction of ectopic eyes by targeted expression of the eyeless gene in *Drosophila*'. *Science* 267, pp. 1788–92.
17 S.A. Kauffman (1969). 'Metabolic stability and epigenesis in randomly constructed nets'. *Journal of Theoretical Biology* 22, pp. 437–67. S.A. Kauffman (1991). 'Antichaos and adaptation'. *Scientific American* 265(2), pp. 64–70. S.A. Kauffman (1993). *The Origins of Order: Self-organization and Selection in Evolution*. Oxford, Oxford University Press.

Chapter 10

1 F. Bisby (2000). 'The quiet revolution: biodiversity informatics and the Internet'. *Science* 289, pp. 2309–312.
2 D.G. Green and T.R.J. Bossomaier (2001). *Online GIS and Spatial Metadata*. Taylor & Francis, London.
3 Open GIS Consortium at http://www.opengis.org.

4 Open GIS Consortium (2000). *Geography Markup Language (GML) v. 1.0.*
5 http://pubweb.parc.xerox.com/map.
6 United Nations Environment Programme (UNEP) (1995). 'Background documents on the Clearing-House Mechanism (CHM)'. *Convention on Biological Diversity*. Jakarta, Indonesia. http://www.biodiv.org/chm/info/official.html.
7 D.G. Green (1994). 'Databasing diversity: a distributed, public-domain approach'. *Taxon* 43, pp. 51–62.
8 United Nations Environment Programme (UNEP) (1992). *Convention on Biological Diversity*. Montreal: Secretariat of the Convention of Biological Diversity.
9 H.M. Burdet (1992). 'What is IOPI?' *Taxon* 41, pp. 390–2. http://life.csu.edu.au/iopi.
10 IUBS (1998). *Species 2000*. International Union of Biological Sciences. http://www.sp2000.org.
11 United Nations Environment Programme (UNEP) (1995). 'Background documents on the Clearing-House Mechanism (CHM)'. *Convention on Biological Diversity*. Jakarta, Indonesia. http://www.biodiv.org/chm/info/official.html.
12 The Environmental Resources Information Network (ERIN) commenced operations in 1990. The environmental information systems that it set up became models for similar projects in many other countries. The systems later became absorbed into the online biodiversity and environmental services provided by Environment Australia (http://www.environment.gov.au).
13 G. Hardy (1998). The OECD's Megascience Forum Biodiversity Informatics Group. http://www.oecd.org//ehs/icgb/BIODIV8.HTM.
14 Two of the most notable systems for predicting species distributions have been BIOCLIM (plants and animals) and CLIMEX (insects). Relevant sources include J.R. Busby (1991). 'BIOCLIM: A bioclimate analysis and prediction system', in C.R. Margules and M.P. Austin (eds), *Nature Conservation: Cost Effective Biological Surveys and Data Analysis*. CSIRO Australia, Collingwood, pp. 64–7; and R.W. Sutherst, G.F. Maywald, T. Yonow and P.M. Stevens (1999). *CLIMEX: Predicting the Effects of Climate on Plants and Animals*. CSIRO Publishing, Collingwood, Australia.

Chapter 11

1 J. McLeod and J. Osborn (1966). 'Physiological simulation in general and particular', in H.H. Pattee, E.A. Edelsack, Louis Fein, A.B. Callahan (eds), *Natural Automata and Useful Simulations*. Spartan Books, Washington DC, pp. 127–38.

2 ActiveWorlds.com (2000). Alphaworld. http://www.activeworlds.com.
3 Virtual Worlds Movement (http://www.virtualworlds.org).
4 See H. Moravec, (1988). *Mind Children: The Future of Robot and Human Intelligence.* Harvard University Press, Cambridge, Mass.
5 For a more detailed description, see T.R.J. Bossomaier and D.G. Green (1998). *Patterns in the Sand: Computers, Complexity and Life.* Allen & Unwin, Sydney.
6 E.F. Loftus and K. Ketcham (1994). *The Myth of Repressed Memory.* St. Martin's Press, New York. See also *Scientific American* 277(3), September 1997, pp. 70–5.
7 Daniel Dennett (1991). *Consciousness Explained.* Penguin, London.
8 W. Gibson (1984). *Neuromancer.* Ace Books, New York.
9 A. Huxley (1962). *The Island.* (First published in 1962. Reissued in 1989.) HarperCollins, New York.
10 M. Wertheim (1999). *The Pearly Gates of Cyberspace.* Doubleday, New York.

Chapter 12

1 Newsgroup: alt.alien.research,sci.astro,sci.math,sci.physics,sci.physics.relativity Subject: Re: Mars GS and Cydonia Date: 29 Mar 1998 18:28:13 GMT.
2 Alvin Toffler (1970). *Future Shock.* Random House, New York.
3 US Bureau of Labor Statistics (1998) (http://stats.bls.gov/ces8mlr.htm).
4 McLuhan was referring to the way radio had brought the world into closer contact. He develops the idea of the global village in several books. M. McLuhan (1964). *Understanding Media.* Mentor, New York; M. McLuhan and Q. Fiore (1967). *The Medium is the Massage.* Bantam, New York; and M. McLuhan and Q. Fiore (1968). *War and Peace in the Global Village.* Bantam, New York.
5 Endangered Languages Repository (http://www.facstaff.bucknell.edu/rbeard/elr).
6 http://marshall.csu.edu.au.
7 At the time of writing, various governments are experimenting with the notion of 'edemocracy' and 'evoting'. See for instance the following online sites: http://www.noie.gov.au/publications/media_releases/2002/Mar2002/online_council.htm; http://www.egov.vic.gov.au/Research/ElectronicDemocracy/voting.htm.
8 Also called I-war (short for 'information war').

INDEX